Uwe Gotzes

Decision Making with Dominance Constraints in Two-Stage Stochastic Integer Programming

VIEWEG+TEUBNER RESEARCH

Stochastic Programming

Editor:

Prof. Dr. Rüdiger Schultz

Uncertainty is a prevailing issue in a growing number of optimization problems in science, engineering, and economics. Stochastic programming offers a flexible methodology for mathematical optimization problems involving uncertain parameters for which probabilistic information is available. This covers model formulation, model analysis, numerical solution methods, and practical implementations. The series "Stochastic Programming" presents original research from this range of topics.

Uwe Gotzes

Decision Making with Dominance Constraints in Two-Stage Stochastic Integer Programming

With a foreword by Prof. Dr. Rüdiger Schultz

VIEWEG+TEUBNER RESEARCH

Bibliografische Information der Deutschen Nationalbibliothek
Die Deutsche Nationalbibliothek verzeichnet diese Publikation in der
Deutschen Nationalbibliografie; detaillierte bibliografische Daten sind im Internet über
<http://dnb.d-nb.de> abrufbar.

Beim vorliegenden Buch handelt es sich um eine vom Fachbereich Mathematik
der Universität Duisburg-Essen genehmigte Dissertation.

Datum der mündlichen Prüfung: 27. März 2009

Referent: Prof. Dr. Rüdiger Schultz
Korreferent: Prof. Dr. Maarten H. van der Vlerk

1. Auflage 2009

Alle Rechte vorbehalten
© Vieweg+Teubner | GWV Fachverlage GmbH, Wiesbaden 2009

Lektorat: Christel A. Roß | Anita Wilke

Vieweg+Teubner ist Teil der Fachverlagsgruppe Springer Science+Business Media.
www.viewegteubner.de

Umschlaggestaltung: KünkelLopka Medienentwicklung, Heidelberg
Gedruckt auf säurefreiem und chlorfrei gebleichtem Papier.
Printed in Germany

ISBN 978-3-8348-0843-1

Foreword

Stochastic dominance, an established concept in decision theory, has gained attention in stochastic programming only recently. The present monograph contributes to this line of research. It deals with stochastic programming models incorporating risk aversion via stochastic dominance constraints. The latter arise by comparing decision dependent random variables and pre-specified benchmarks. This induces some notion of acceptance: Only those decisions are feasible that lead to random entities, e. g., costs, returns, or revenues, which compare favorably to some random benchmark profile reflecting the user's desire.

This monograph addresses decision making with stochastic dominance constraints in the framework of two-stage mixed-integer linear stochastic programming. Its main results concern basic structural findings, novel decomposition algorithms for the numerical solution of the large-scale stochastic programs arising, and case studies on two exemplary industrial optimization problems under uncertainty, namely competitive selling price determination for electricity retailers and planning of a local network for heat supply. Thus, both readers interested in mathematical foundation or practical application of optimization under uncertainty may find this text interesting.

The monograph grew out of a doctoral dissertation prepared during 2005–2008 at the Chair of Discrete Mathematics and Optimization in the Department of Mathematics of the University of Duisburg-Essen. This research has been supported by the German Federal Ministry of Education and Research (BMBF) within the program "Netzwerke Grundlagenforschung erneuerbare Energien und rationelle Energieanwendung".

Rüdiger Schultz

Acknowledgements

First, I would like to thank my supervisor Rüdiger Schultz for his invaluable guidance and continuous support. Rüdiger was always there to listen and to give advice. He is responsible for involving me in the interdisciplinary BMBF[1] funded project "Dezentrale regenerative Energieversorgung: Innovative Modellierung und Optimierung"[2] in the first place, which made this thesis possible.

My grateful thanks also go to Ralf Gollmer and Harald Held for fruitful discussions and especially for bringing me up to date in the Linux world.

Cordial thanks to Frederike Neise and Ralf Gollmer for close collaboration during the modification of Andreas Märkert's decomposition algorithm [75, 76] to our special needs.

Many thanks also to Miguel Carrión for amicable collaboration when we were working on the paper [25], on which Chapter 3 of this thesis is based upon.

Further I am grateful to Mark Lutter for encouraging motivating words and to Florian Liehr and John Klinkhammer for hours and hours of discussions on mathematics during my studies. Without their support things would surely be different.

Uwe Gotzes

---

[1] German Federal Ministry of Education and Research
[2] Distributed renewable power generation: Innovative modeling and optimization

Contents

Chapter 1

Introduction

This work deals with *Stochastic Programming*. Uncertainty is a key issue in many decision problems and ignoring randomness easily leads to inferior or even infeasible decisions. In contrast to the neighboring mathematical fields, such as online or robust optimization [3, 15, 16], stochastic programming models benefit from the assumption that probability distributions governing the data are known. This underlying probabilistic model of uncertainty turns finding optimal decisions into selecting "best" random variables and evokes the need to adequately compare random variables according to their utility in the respective context.

Weighting possible outcomes via statistical parameters, considering probabilities of events of interest or imposing stochastic orders on sets of random variables brings up natural definitions of objectives and constraints in mathematical programming models involving stochastic data.

Dynamic programming and stochastic control, are two fields where stochastic optimization has been analyzed over continuous time [26, 47], as in the physical world, dynamical systems are naturally described in this domain.

However, most system identification schemes have been based on discrete-time models. In particular, there exists an extensive theory and methodology to treat randomness in linear, linear mixed-integer and nonlinear programming models [12, 88, 100, 108]. Since the 1950s and the early works of Beale [13] and Dantzig [32] that initiated stochastic programming with recourse—nowadays a sophisticated branch of stochastic programming—there is a growing community of mathematicians, computer scientists, economists and people from other disciplines advancing the knowledge of how to measure, formalize, model and manage uncertainty in optimization problems. For example, recently ideas from two-stage stochastic programming merged into level-set-based shape optimization [31].

In this introductory chapter we will review some mainstreams addressing the presence of random data in optimization problems. The subsequent chapters will deal with an innovative means to incorporate risk aversion into recourse models. In Chapter 2, *stochastic integer programs with increasing convex order constraints* are introduced. These are the objects we will focus on throughout this work. As a basic qualitative stability result, closedness of the constraint set mapping with respect to perturbations of the underlying probability measure is derived. For discrete probability measures, large-scale, block-structured, mixed-integer linear programming equivalents to the stochastic programs with increasing convex order constraints are identified. Chapter 3 presents results obtained from the application of standard software to deterministic equivalents of a real-life problem from energy trading. In Chapter 4, we will show how the special problem structure can be exploited for an alternative algorithmic treatment of the model. Computational tests with our C-implementation [54, 67] of the ideas from Chapter 2 and instances from power optimization and Sudoku puzzling follow in Chapter 5. Chapter 6 features yet another mixed-integer linear programming formulation for increasing convex order constrained models. Similarities and differences to the formulation from Chapter 2 are pointed out and comparative computational results are discussed. Parts of this study have been published in [25, 49, 50, 51, 54, 55].

An optimization model with uncertain data in a general form is given by:

$$\min_{x \in \mathscr{D}} \{ f(x, \mathsf{Y}) : g(x, \mathsf{Y}) \leq 0 \}. \tag{1.1}$$

Here $f : \mathscr{D} \times \Xi \to \mathbb{R}$ is the objective function and $g : \mathscr{D} \times \Xi \to \mathbb{R}^m$ is a vector of constraint functions. $\emptyset \neq \mathscr{D} \subset \mathbb{R}^n$ denotes a set of deterministic constraints and $\mathsf{Y} : \Omega \to \Xi \subset \mathbb{R}^s$ is a random vector on a probability space $(\Omega, \mathscr{F}, \mathbb{P})$.[3] $\mathbb{R}^s$ carries the Borel σ-algebra $\mathscr{B}^s$.

Immediately the question for the meaning of feasibility and optimality arises because it is not clear that f attains its infimum and if so it might be heavily dependent on the outcome of Y. Obviously, the latter also holds accordingly for the constraint mapping g.

At first let f, g be lower semicontinuous and $\mathscr{D}$, Ξ be closed. A rough straightforward approach to overcome the mentioned ambiguity in (1.1) is to substitute f by $f' := \mathbb{E} \circ f$ (optimization on average)[4] and the component functions of g by $g_i' := \mathbb{E} \circ g_i$, $i = 1, \ldots, m$ (feasibility on average). The general aim, when being faced with a model of the form (1.1) is to find an in a way optimal *here-and-now decision* before knowing the actual outcome of Y.

---

[3]For a brief discussion of basic concepts and definitions from probability theory we refer to the appendices of [89] and [100].

[4]$\mathbb{E}$ denotes the *expected value operator*—there is a symbol index at the end of the text to look up symbols.

1.1 Two-Stage Stochastic Mixed-Integer Linear Programs with Recourse

As already mentioned, two-stage linear programs are among the most widely used and analyzed stochastic programming models. Here the decision maker takes some action in the first stage, after which a random event constitutes the actual data of the model. In the second stage a recourse decision that, for instance, compensates for bad effects that might have been experienced as a result of the first-stage decision can be made.

In many cases there is a natural two-stage framework given through the modelling background and the recourse actions are more just decisions in the future than decisions that fix deficiencies induced by former decisions. Due to this modelling background we prefer to speak of second-stage decisions rather than of recourse decisions. The solution of such a model is a single first-stage decision and a collection of second-stage policies defining which actions should be taken in response to each random outcome. To be more specific, as a first representation of (1.1), let us consider the following mixed-integer linear program involving a random vector $z : \Omega \mapsto \mathbb{R}^s$.

$$\min\left\{ c^\top x + q^\top y : Tx + Wy = z,\ x \in X,\ y \in \mathbb{Z}_+^{\bar{m}} \times \mathbb{R}_+^{m'} \right\}, \qquad (1.2)$$

together with the *information constraint* that

$$x \text{ must be selected prior to observing } z(\omega).$$

Afterwards, in a second stage, when the actual outcome of z is known, the decision on $y = y(x, z(\omega))$ has to be taken. This condition often is referred to as *nonanticipativity* of x. Furthermore, as another basic assumption it is claimed that the outcome of z is independent of the decision on x. For ease of presentation we consider stochasticity only in the right-hand side, but uncertainty may also be present in c, q, T and W. We assume that the vectors c and q as well as the matrices T and W in (1.2) have compliant dimensions, that W has solely rational entries, and that $X \subset \mathbb{R}^m$ is a nonempty polyhedron, possibly involving integer requirements to components of x.

In traditional two-stage stochastic programming, see [20, 65, 94, 100], the aim is to optimize first-stage decisions. To this end, well-defined optimization problems in x and y, often called *deterministic equivalents*, are formulated. The principal construction is as follows. Rewrite (1.2) as

$$\inf_x\left\{ c^\top x \;+\; \inf_y\{ q^\top y : Wy = z - Tx,\ y \in \mathbb{Z}_+^{\bar{m}} \times \mathbb{R}_+^{m'} \} : x \in X \right\} \qquad (1.3)$$

$$= \inf_x\left\{ c^\top x \;+\; \Phi(z - Tx) : x \in X \right\}, \qquad (1.4)$$

where

$$\Phi : \mathbb{R}^s \ni t \longmapsto \inf\{q^\top y \, : \, Wy = t, \, y \in \mathbb{Z}_+^{\bar{m}} \times \mathbb{R}_+^{m'}\} \in \mathbb{R} \cup \{\pm\infty\}. \tag{1.5}$$

One possibility to look at (1.4) is to recognize a family of random variables

$$(f_x \circ z)_{x \in X}, \text{ with} \tag{1.6}$$

$$f_x : \mathbb{R}^s \ni u \longmapsto c^\top x + \Phi(u - Tx) \in \mathbb{R} \cup \{\pm\infty\}, \tag{1.7}$$

and to understand (1.4) as the problem of finding a "best" member in the family (1.6). For the sake of convenience we introduce the random variable

$$\tilde{f}_x : \Omega \ni \omega \longmapsto (f_x \circ z)(\omega) \in \mathbb{R} \cup \{\pm\infty\}, \tag{1.8}$$

which is $\mathscr{F} - \overline{\mathscr{B}}^1$-measurable because it can be understood as a composition of z and Φ with continuous mappings $g : \mathbb{R} \to \mathbb{R}$ and $h : \mathbb{R}^s \to \mathbb{R}^s$: $\tilde{f}_x = g \circ \Phi \circ h \circ z$. As we will see later in this section, Φ is lower semicontinuous (and real-valued) under mild assumptions. The lower level sets $\{v \in \mathbb{R}^s \, : \, \Phi(v) \leq \alpha, \, \alpha \in \mathbb{R}\}$ are closed then and thus are contained in $\mathscr{B}^s$ which is a measurability criterion.

The most straightforward way to make the selection for x in (1.6) is to compare the random variables by their expectations, leading to the deterministic equivalent

$$\min\{\mathbb{E}[\tilde{f}_x] \, : \, x \in X\}. \tag{1.9}$$

In the literature this problem is known as the (classical) two-stage stochastic program with (mixed-integer) linear recourse, [20, 65, 94, 100].

Opposed to the purely linear case, the rationality requirement on W is necessary to conclude solvability of the second-stage problem from its feasibility and boundedness. For instance the integer problem $\min\{\sqrt{2}x_1 - x_2 \, : \, \sqrt{2}x_1 - x_2 \geq 0, \, x_1 \geq 1, \, x_1, x_2 \in \mathbb{Z}_+\}$ is feasible and bounded (the infimum is 0)[5], but an optimal solution does not exist, because $\sqrt{2}x_1 - x_2 = 0$ is a contradiction to the irrationality of $\sqrt{2}$.

The purely expectation-based, risk neutral problem (1.9) can be extended to a model involving risk aversion if the random variables $\tilde{f}_x$ get compared via statistical parameters reflecting risk. With a measure of risk $\mathscr{R}$ and a fixed weight factor $\rho > 0$, the mean-risk extension of (1.9) reads

$$\min\{\mathbb{E}[\tilde{f}_x] + \rho \cdot \mathscr{R}[\tilde{f}_x] \, : \, x \in X\}. \tag{1.10}$$

---

[5]Among others, this can be obtained from Dirichlet's pigeon hole principle, cf. [39, 101]. Dirichlet proved that for $\alpha \in \mathbb{R}$ and $\varepsilon \in (0, 1]$ there exist integers p, q such that $\mid \alpha - \frac{p}{q} \mid < \frac{\varepsilon}{q}$ and $1 \leq q \leq \varepsilon^{-1}$ (approximation of irrational numbers by rational numbers with "small" denominators).

Risk measures $\mathscr{R}$, leading to mixed-integer linear programming equivalents with desirable characteristics that were used in this context include both, quantile-based (Excess Probability, Value at Risk, Conditional Value at Risk) and deviation-based[6] measures (Expected Excess, Semideviation), see [2, 43, 70, 77, 95, 96, 103, 104, 105]. In their paper [8], Artzner et al. outline, in axiomatic fashion, the nowadays widely accepted properties a risk measure[7] should possess[8] to be considered *coherent*. Note that the expected value operator is a linear functional and hence defines a coherent risk measure. Artzner et al. proposed the *Worst Conditional Expectation* as a coherent measure of risk [8]. The *Value at Risk* is a widespread measure of risk applied in finance for quantitative risk management for many types of risk. However it does not fulfill the subadditivity axiom, saying that "a merger does not create additional risk", cf. [8]. In [8] four examples illustrate why subadditivity is a natural requirement. This shortcoming might be the reason for a gain in popularity of the *Conditional Value at Risk* in mathematical finance, being a coherent measure, cf. [1, 8, 90, 96]. The above mentioned measures *Expected Excess* and *Excess Probability*, due to fixed cost targets, do not fulfill the coherency axioms. But these measures fulfill the *coherency axioms with respect to fixed targets* proposed in [75]. The *Semideviation* is neither translation invariant nor monotonic. However risk measures resulting from a compound of $\mathbb{E}$ and $\mathscr{R}$ none the less might be coherent for certain values of the risk parameter ρ, cf. [85, 75]. The *Variance* as a commonly used measure of the spread of the values of a random variable is in our context in many respects unfavorable. For instance, it is not granted that the mean-risk model (1.10), with the variance as risk-functional is well posed in the sense that its infimum exists and is attained, provided that $X \neq \emptyset$ and compact.

For the structure of (1.9) and (1.10) the second-stage value function Φ from (1.5) is of fundamental importance. To ensure that Φ is finite for all $t \in \mathbb{R}^s$, we assume

$$
\begin{aligned}
&\text{(A0)} && W \in \mathbb{Q}^{s \times (\bar{m}+m')} && \text{(rationality)} \\
&\text{(A1)} && W\left(\mathbb{Z}_+^{\bar{m}} \times \mathbb{R}_+^{m'}\right) = \mathbb{R}^s && \text{(complete recourse)} \\
&\text{(A2)} && \left\{u \in \mathbb{R}^s : W^\top u \leq q\right\} \neq \emptyset && \text{(sufficiently expensive recourse)}.
\end{aligned}
$$

(A1) can generally be relaxed to the claim that $W\left(\mathbb{Z}_+^{\bar{m}} \times \mathbb{R}_+^{m'}\right) \supset \{z(\omega) - Tx : \omega \in \Omega, x \in X\}$ (relatively complete recourse). (A2) postulates the dual feasibility of

---

[6] measures that depend on the expected deviation of the random variable from some target

[7] In the mentioned paper, measures of risk are mappings on the set of real random variables on *finite* probability spaces. Subsequently Delbaen extended the theory to general probability spaces [34].

[8] in detail: Translation invariance, subadditivity, positive homogeneity and monotonicity

the LP relaxation of the minimum problem behind Φ. For proofs of the following statements we refer to [11, 21].

Theorem 1.1. *Suppose (A0)–(A2). Then it holds:*

(1) Φ *is real-valued and lower semicontinuous on* $\mathbb{R}^s$, *i. e.* $\liminf_{t \to t'} \Phi(t) \geq \Phi(t')$ *for all* $t' \in \mathbb{R}^s$.

(2) *There exists a countable partition* $\bigcup_{i=1}^{\infty} \mathcal{T}_i = \mathbb{R}^s$ *such that the restrictions of* Φ *to* $\mathcal{T}_i$ *are piecewise linear and Lipschitz continuous with a uniform constant* $L > 0$ *not depending on i.*

(3) *Each of the sets* $\mathcal{T}_i$ *has a representation* $\mathcal{T}_i = \{\{t_i\} \oplus \mathcal{K}\} \setminus \bigcup_{j=1}^{N} \{\{t_{ij}\} \oplus \mathcal{K}\}$, *where* $\mathcal{K}$ *denotes the polyhedral cone*

$$\left\{ W \begin{pmatrix} \bar{0}_{\bar{m}} \\ y' \end{pmatrix} : y' \in \mathbb{R}_+^{m'} \right\}.$$

$\bar{0}_{\bar{m}}$ *denotes an* $\bar{m}$*-dimensional null vector. Moreover N does not depend on i.*

(4) $\exists \alpha, \beta > 0$ *such that* $|\Phi(t_1) - \Phi(t_2)| \leq \alpha \|t_1 - t_2\| + \beta$, $\forall\, t_1, t_2 \in \mathbb{R}^s$.

Walkup and Wets proved a basis decomposition theorem revealing the structure of Φ in the case of linear recourse in [115]. There Φ has the objective function vector as a second argument. Without integer requirements in the second stage linear programming duality together with (A1) and (A2) yields

$$\begin{aligned}
\Phi(t) &= \min\left\{ q^\top y : Wy = t, y \geq 0 \right\} \\
&= \max\left\{ t^\top u : W^\top u \leq q \right\} \\
&= \max_{\ell=1,\dots,L} d_\ell^\top t,
\end{aligned} \tag{1.11}$$

where d_ℓ, $\ell = 1,\dots,L$ are the vertices of $\{u : W^\top u \leq q\}$. Hence, Φ, as a pointwise maximum of finitely many linear functions, is piecewise linear and convex in this case.

Theorem 1.1 shows that two main impacts of integrality restrictions in (1.5) concern the convexity and the continuity properties of Φ (see [102] for illustrative examples).

Based on Theorem 1.1 it could be shown that the mean-risk models using the risk-measures Excess Probability, Expected Excess, Value at Risk, and Conditional Value at Risk do not suffer from ill-posedness drawbacks as for instance the mean-variance approach [77, 104, 105, 110]. The just mentioned publications also provide results concerned with stability aspects, i.e., results concerning certain continuity properties of the *optimal value function* and the *solution set mapping* (see [97], also containing an extensive bibliography on the topic, for a recent overview on stability analysis in stochastic programming).

The two-stage model can be seen as a special case of a more general problem class, called multistage stochastic programming models. In these models, the decision variables and constraints are divided into stages $t = 1, \ldots, T$. Again, the information structure, now typically represented by a discrete time stochastic process and more complex nonanticipativity restrictions, commonly formulated as measurability conditions on the stochastic decision variables, is fundamental: What is known at which stage, when decisions are made in the multiperiod framework (cf. [42, 98])? Early contributions to the multistage approach came up in Beale et al. [14], Louveaux [72] and Birge [19]. As in two-stage stochastic programming, in the traditional setting the expectation of a suitable term is used as objective [20, 65, 94]. More recently risk aversion also became an important issue in multistage models (cf. [43, 57] and the references therein).

Another major approach to account for randomness in optimization problems is *chance constrained programming.* In contrast to two- and multistage models, where the compensating costs are assumed to be known for all scenarios, chance constrained programming can be useful when constraint violation cannot be avoided because of unexpected extreme events in all cases or when compensations do not exist or cannot be expressed in monetary form. Feasibility "as much as possible" is the fundamental idea behind chance constrained programming [93, 94].

1.2 Risk Aversion by Stochastic Ordering Constraints

In Section 3.6 of their famous book *Theory of Games and Economic Behavior* [114], which introduced the tools of modern logic into economics and laid out the theoretical foundations of *neoclassical economics* (cf. [113]), von Neumann and Morgenstern formulated a set of axioms for a rational decision maker. These axioms, each of them having an intuitive meaning ([114, Section 3.6.2]), imply the so-called *expected utility hypothesis.* It says that for a rational decision maker there exists a utility function u such that the random variable X is preferred to the random variable Y iff $\mathbb{E}[u(X)] \geq \mathbb{E}[u(Y)]$. In practice, it turns out that it is almost impossible to identify the utility function of a decision maker explicitly (see also [63]). However, one could wonder, whether there are mathematical characteristics for the distributions of X and Y that allow for prediction of the decision maker's choice when there is only partial knowledge of its utility function, say, that it belongs to some class $\mathscr{G}$ of functions (cf. [79]). If the decision maker is rational in the sense of preferring more to less, then the utility function is necessarily nondecreasing. We call a decision maker risk averse, if the certain yield of $\mathbb{E}(X)$ is preferred to every risky/stochastic outcome $X(\omega)$, i. e., the utility function satisfies $\mathbb{E}[u(X)] \le u[\mathbb{E}(X)]$ for all X. Jensen has shown that this holds, iff u is concave.[9] Hence, a rational (risk averse) decision maker will prefer X to Y iff $\mathbb{E}[u(X)] \ge \mathbb{E}[u(Y)]$ for all nondecreasing (concave) $u : \mathbb{R} \to \mathbb{R}$. This leads us to the following basic definition.

Definition 1.2. Let X ,Y be real random variables with finite means. Then we say

(i) X dominates Y by the *first degree stochastic dominance* (FSD) rules, which we write as $X \succeq_1 Y$, iff

$$\mathbb{E}[f \circ X] \ge \mathbb{E}[f \circ Y] \ \forall \text{ nondecreasing } f : \mathbb{R} \to \mathbb{R}$$

for which both expectations exist.

(ii) X dominates Y by the *second degree stochastic dominance* (SSD) rules, which we write as $X \succeq_2 Y$, iff

$$\mathbb{E}[f \circ X] \ge \mathbb{E}[f \circ Y] \ \forall \text{ nondecreasing and concave } f : \mathbb{R} \to \mathbb{R}$$

for which both expectations exist.[10]

By choosing the above relations as definitions for first- and second degree (also: first- and second order) stochastic dominance, it is clear that $\succeq_i$, $i = 1,2$ belong to the so called *integral stochastic orders* ([117]). These stochastic orders can be characterized by a class $\mathscr{G}$ of functions: $X \succeq Y$ iff $\mathbb{E}[f \circ X] \le \mathbb{E}[f \circ Y] \ \forall f \in \mathscr{G}$.

---

[9]Jensen (1906): Let $(\Omega, \mathscr{A}, \mu)$ be a measure space, μ a probability measure, $I \subset \mathbb{R}$ an interval, $f : \Omega \to I$ μ-integrable and $\varphi : I \to \mathbb{R}$ convex (concave). Then, $\int_\Omega f d\mu \in I$, $\varphi \circ f$ is quasiintegrable and it holds: $\varphi(\int_\Omega f d\mu) \le (\ge) \int_\Omega \varphi \circ f d\mu$ ([44, 62]). The other inclusion follows from choosing a measurable set $A \subset \Omega$ with $\mu(A) = \lambda$ and the family $X_{x,y}(\omega) := \mathbb{1}_A(\omega) \cdot x + \mathbb{1}_{\Omega \setminus A}(\omega) \cdot y$, $\forall \omega \in \Omega$.

[10]Generalizations of variability orders such as FSD and SSD to random variables with non-existing, or infinite means lead to serious difficulties, see, for example [45]. Note the one-to-one correspondence $\int f \circ X \, d\mathbb{P} \le \int f \circ Y \, d\mathbb{P} \Leftrightarrow \int f \, d\mathbb{P}_X \le \int f \, d\mathbb{P}_Y$ (integration with respect to the image measure). In [45] the ordering $\succeq$, with $Q \succeq P :\Leftrightarrow Q \in \{R$ Borel probability measure on $M : R$ can be obtained from P by collapsing parts of the mass of P to their respective barycenters $\}$ (this set contains the so called fusions of P on a (e. g., separable Banach) space M) is introduced. It is shown that $Q \succeq P$ implies $Q \succeq_{cx} P(:\Leftrightarrow Q \succeq_2 P \wedge \int dQ = \int dP)$. The crucial point is, that $\succeq$ is in general *not antisymmetric* if Q or P do not have finite means.

The term "integral stochastic order" refers to the integration process behind the expectation operator. $\mathscr{G}$ is called a generator of the stochastic order.

Stochastic dominance, having gained some attraction in stochastic programming in recent years [35, 36, 37], is one of the fundamental concepts to answer the question how deciders make or should make decisions and how optimal decisions can be reached (cf. [71, 116]).

Since it is impossible to verify the inequality $\mathbb{E}[f \circ X] \geq \mathbb{E}[f \circ Y]$ for *all* non-decreasing (concave) functions. The next results show that it is sufficient to point-wisely compare *two* performance functions for the verification of the dominance relations, namely the distribution functions F_X and F_Y (Proposition 1.3, (i)) and the integrated distribution functions of the involved random variables (Proposition 1.3, (ii) together with Lemma 1.4, (i)).

Proposition 1.3.

> (i) $X \succeq_1 Y \iff F_X(t) \leq F_Y(t), \ \forall t \in \mathbb{R}.$

> (ii) $X \succeq_2 Y \iff \mathbb{E}[(t - X)_+] \leq \mathbb{E}[(t - Y)_+], \ \forall t \in \mathbb{R}.$

The expression $(.)_+$ is defined as $\max\{.,0\}$. In view of Definition 1.2, (ii), note that $\mathbb{E}[(t - X)_+] \leq \mathbb{E}[(t - Y)_+] \Leftrightarrow \mathbb{E}[-((t - X)_+)] \geq \mathbb{E}[-((t - Y)_+)]$ and that $f(.) := -((t - .)_+)$ is nondecreasing and concave.

For a proof of 1.3 (i) see Theorem 1.2.8 and Definition 1.2.1 in [79]. 1.3 (ii) shows that the generator of SSD can be significantly thinned out. For a proof, [79] is again a good reference (Theorem 1.5.7).[11] Here SSD, a term occurring in the theory of decision under risk, is called *increasing concave order* which is closer to the definition. FSD is called *(usual) stochastic order*, as it is the most natural candidate for a stochastic order. The next theorem gives an equivalent representation of the expressions $\mathbb{E}\left[(t - X)_+\right]$ and $\mathbb{E}\left[(X - t)_+\right]$ which will turn out useful later.

Lemma 1.4. *It holds*

> (i) $\mathbb{E}\left[(t - X)_+\right] = \int_{-\infty}^{t} F_X(z)\, dz.$

> (ii) $\mathbb{E}\left[(X - t)_+\right] = \int_{t}^{\infty} 1 - F_X(z)\, dz.$

A trivial conclusion from Definition 1.2 is that $X \succeq_1 Y \Rightarrow X \succeq_2 Y$, since the generator of $\succeq_2$ is a subset of $\succeq_1$'s generator. This can also easily be seen from

---

[11] The assertion is shown for the *increasing convex order* there. Note that X is less than Y in increasing concave order iff $-Y$ is less than $-X$ in increasing convex order. Applying the theorem we have equivalence to $\mathbb{E}[(-Y - t)_+] \leq \mathbb{E}[(-X - t)_+] \ \forall t \in \mathbb{R}$ which in turn is equivalent to $\mathbb{E}[(t - Y)_+] \leq \mathbb{E}[(t - X)_+] \ \forall t \in \mathbb{R}$ (replace t by $-t$).

Proposition 1.3 together with Lemma 1.4 (i). Obviously $X \succeq_1 Y \Rightarrow X \succeq_2 Y$ does not hold in general, but there is a special case where SSD and FSD are equivalent:

Corollary 1.5. $(X \succeq_2 Y \wedge |Y(\Omega)| = 1) \implies X \succeq_1 Y.$

Proof. There exists a ξ^* such that

$$F_Y(\xi) = \begin{cases} 0, & \text{if } \xi < \xi^* \\ 1, & \text{otherwise.} \end{cases}$$

Assume $\exists\, \tilde{\xi} \in \mathbb{R} : F_X(\tilde{\xi}) > F_Y(\tilde{\xi})$. Then $\tilde{\xi} < \xi^*$ holds, since $0 \leq F_X \leq 1$. Hence $0 = \int_{-\infty}^{\xi^*} F_Y(z)\,dz \geq \int_{-\infty}^{\xi^*} F_X(z)\,dz \implies F_X(\tilde{\xi}) = 0$ (monotonicity). This yields a contradiction to $F_X(\tilde{\xi}) > F_Y(\tilde{\xi})$. $\qquad\square$

In the case where Y only has one realization the question, whether X dominates Y to first- or second order simplifies to the question if X falls below a fixed critical value ξ^* with a positive probability or not.

By choosing $f := id_{\mathbb{R}}$ in Definition 1.2, it is clear that $X \succeq_i Y \Rightarrow \mathbb{E}(X) \geq \mathbb{E}(Y)$, $i = 1, 2$ and we say that the expectation is *consistent* with first and second order stochastic dominance. More general, a mean-risk model is *α-consistent with FSD/SSD* iff $X \succeq_i Y \Rightarrow \mathbb{E}(X) + \alpha \cdot \mathscr{R}(X) \geq \mathbb{E}(Y) + \alpha \cdot \mathscr{R}(Y)$ for $i = 1$ or $i = 2$, respectively (see [46, 75, 83, 84] for consistency analysis).

Next we will adapt Definition 1.2 to a minimization framework. It can easily be shown that

$$-X \;\succeq_1\; -Y \quad\Longleftrightarrow\quad Y \;\succeq_1\; X$$
$$\overset{(\text{Pr.1.3})}{\Longleftrightarrow} \quad F_X(t) \;\geq\; F_Y(t) \quad \forall t \in \mathbb{R}.$$

For second order stochastic dominance easy manipulations (use Proposition 1.3 and replace t by $-t$) yield

$$-X \;\succeq_2\; -Y$$
$$\Longleftrightarrow \quad \mathbb{E}\left[(X - t)_+\right] \;\leq\; \mathbb{E}\left[(Y - t)_+\right] \quad \forall t \in \mathbb{R}.$$

Taking [79, Theorem 1.5.7] into account, Definition 1.2 reflecting preference of smaller instead of larger outcomes reads:

Definition 1.6. Let X, Y be real random variables with finite means. Then we say

  (i) X is better than Y, with respect to the *first degree stochastic dominance* rules when *preferring smaller outcomes*, which we write as $X \leq_{st} Y$, iff

$$\mathbb{E}[f \circ X] \leq \mathbb{E}[f \circ Y] \;\; \forall \text{ nondecreasing } f : \mathbb{R} \to \mathbb{R}$$

    for which both expectations exist.

(ii) X is better than Y, with respect to the *second degree stochastic dominance rules* when *preferring smaller outcomes*, which we write as $X \leq_{icx} Y$, iff

$$\mathbb{E}[f \circ X] \leq \mathbb{E}[f \circ Y] \ \forall \text{ nondecreasing and convex } f : \mathbb{R} \to \mathbb{R}$$

for which both expectations exist.

Remark 1.7. "$\leq_{st}$" and "$\leq_{icx}$" are standard notations for the already mentioned *usual stochastic order* and the *increasing convex order*. "$\leq_{icx}$" is the counterpart to the *increasing concave*, or *SSD* order.

To put Definition 1.6 into words, it can be said that $X \leq_{st} Y$ means that X takes on smaller values with no smaller probability than Y. $X \leq_{icx} Y$ means that the expected excess of X above t is not larger than the expected excess of Y above t for any real t. Figure 1.1 shows such a situation for distributions that intersect each other only once. In this case it has to hold that the light gray area has to be less than or equal to the dark gray area. In general it can be said that each area enclosed below F_Y and above F_X "on the left" has to appear enclosed above F_Y and below F_X "on the right". For $X \leq_{st} Y$, F_X has to be pointwise less than or equal to F_Y on the entire line. $\mathbb{P}(\{X > t\}) = 1 - F_X(t)$ is called the survival function of X. Proposition 1.3 (ii) together with Lemma 1.4 (ii) yields a useful criterion for $X \leq_{icx} Y$ using integrated survival functions.

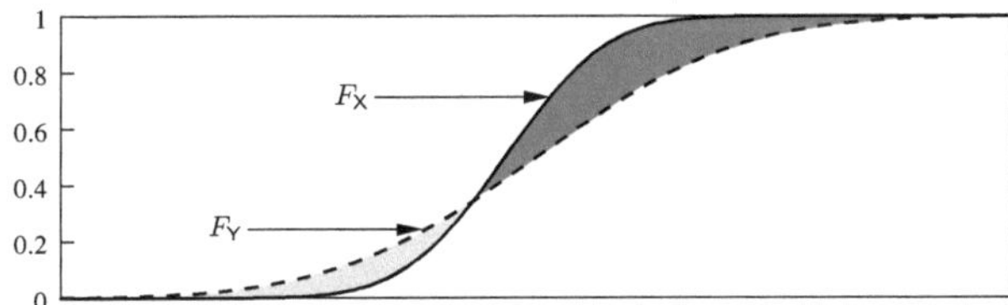

Figure 1.1: This figure reflects $X \leq_{icx} Y$ in terms of distribution functions.

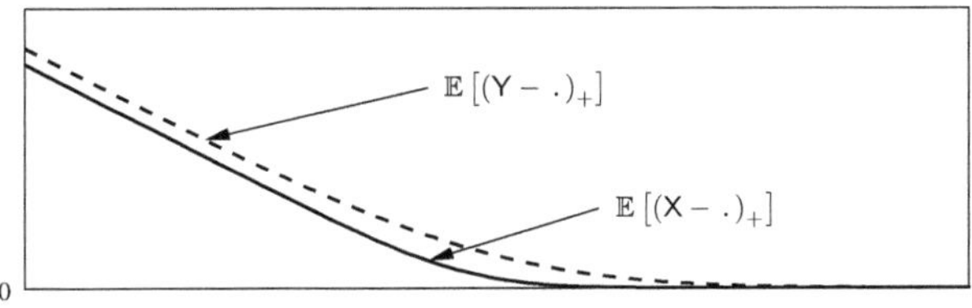

Figure 1.2: $X \leq_{icx} Y$ in terms of the performance functions $\mathbb{E}\left[(X - .)_+\right]$ and $\mathbb{E}\left[(Y - .)_+\right]$.

Remark 1.8. In [83], among others, the *Outcome-Risk (O-R) diagram*, a graphical tool for the (second order) stochastic dominance methodology is introduced. It contains the graphs of $\mathbb{E}[(t-X)_+]$ and $(t-\mathbb{E}[X])_+$ (as functions of t). The graph of $\mathbb{E}[(t-X)_+]$ has two asymptotes, the t-axis on the left and the line $(t-\mathbb{E}[X])$ on the right. The asymptotes intersect at the point $(\mathbb{E}[X], 0)$. Hence, $(t-\mathbb{E}[X])_+$ is the pointwise maximum of the two asymptotes and coincides with $\mathbb{E}[(t-X)_+]$ if X is deterministic. Any uncertain outcome with mean $\mathbb{E}[X]$ yields a graph not below the asymptotes. The space between the curves in the O-R diagram represents the dispersion of X in comparison to $\mathbb{E}[X]$ and is called the *dispersion space*.

Size and shape of the dispersion space are analyzed and related to size parameters summarizing characteristics of riskiness (absolute (semi-)deviation, standard (semi-)deviation).

The results obtained in [83] can easily be transferred to the situation of preferring smaller instead of larger outcomes. For instance, if $\mathbb{E}[X^2] < \infty$, the maximal vertical diameter of the dispersion space is equal to the absolute semideviation[12] of X. Moreover,

$$\int_{-\infty}^{\infty} \mathbb{E}[(X-t)_+] - (\mathbb{E}[X]-t)_+ \, dt = \frac{1}{2}\sigma_X^2$$

and thus if $\mathbb{E}[X^2] = \mathbb{E}[Y^2] < \infty$, the difference of the areas of the dispersion spaces equals one half of the difference of the variances[13] of X and Y:

$$\int_{-\infty}^{\infty} \mathbb{E}[(X-t)_+] - \mathbb{E}[(Y-t)_+] \, dt = \frac{1}{2}(\sigma_X^2 - \sigma_Y^2).$$

---

[12] Absolute semideviation of X (preference of small outcomes): $\mathbb{E}[(X-\mathbb{E}[X])_+]$
[13] Variance: $\sigma_X^2 := \mathbb{E}[(X-\mathbb{E}[X])^2]$

Chapter 2

Increasing Convex Order Constraints Induced by Mixed-Integer Linear Recourse

The starting point of the investigations in this chapter is to identify "acceptable" members of $(\tilde{f}_x)_{x\in X}$ (cf. (1.6), (1.8)) with respect to the introduced partial orders rather than looking for a "best" among them with respect to some scalarization. We assume that a random variable $a : \Omega \to \mathbb{R}$ reflecting an acceptance threshold for the costs $\tilde{f}_x$ resulting from the two-stage dynamics in (1.2) is given. We will consider $x \in X$ acceptable iff $\tilde{f}_x \leq_{icx} a$. Over all acceptable x we minimize an objective function $g : \mathbb{R}^m \to \mathbb{R}$. This leads to the following stochastic program with increasing convex order constraints induced by mixed-integer linear recourse

$$\min \left\{ g(x) \ : \ \tilde{f}_x \leq_{icx} a \, , \, x \in X \right\}. \tag{2.1}$$

This model is closely related to the counterpart model where "$\leq_{icx}$" is replaced by the usual stochastic order, see [52, 80] for an analysis of the latter. Since the increasing convex order is the weaker notion, (2.1) is a relaxation of the model with the usual stochastic order instead of "$\leq_{icx}$", see [82] for related work.

2.1 Structural Properties

The aim of this section is to provide a framework such that the objects in (2.1) are well-defined, and to derive some basic structural properties of (2.1).

Proposition 1.3, (ii) implies, that the defining relation for being smaller in the increasing convex order (1.6, (ii)) is already valid if it holds for all *wedge functions* of the type $h(.) := \max\{(. - \eta), 0\} =: (. - \eta)_+ , \eta \in \mathbb{R}$. Let $\mathscr{P}(\mathbb{R}^s), \mathscr{P}(\mathbb{R})$ be the sets of all Borel probability measures on $\mathbb{R}^s$ and $\mathbb{R}$, and let $\mu := \mathbb{P} \circ z^{-1} \in \mathscr{P}(\mathbb{R}^s)$ and $\nu := \mathbb{P} \circ a^{-1} \in \mathscr{P}(\mathbb{R})$ denote the image measures of $\mathbb{P}$ under z and a on $\mathbb{R}^s$ and $\mathbb{R}$, respectively. The constraint $\tilde{f}_x \leq_{icx} a$ now can be equivalently expressed as

$$\mathbb{E}\left[(\tilde{f}_x - \eta)_+ \right] \ \leq \ \mathbb{E}\left[(a - \eta)_+ \right] \quad \forall \eta \in \mathbb{R} \tag{2.2}$$

$$\Leftrightarrow \quad \int_\Omega (\tilde{f}_x(\omega) - \eta)_+ \, d\mathbb{P} \ \leq \ \int_\Omega (a(\omega) - \eta)_+ \, d\mathbb{P} \quad \forall \eta \in \mathbb{R} \tag{2.3}$$

$$\Leftrightarrow \quad \int_{\mathbb{R}^s} [f_x(\zeta) - \eta]_+ \, \mu(d\zeta) \; \leq \; \int_{\mathbb{R}} [\alpha - \eta]_+ \, \nu(d\alpha) \quad \forall \eta \in \mathbb{R}, \qquad (2.4)$$

where we switched over to integration with respect to the image measures μ and ν and consequently from $\tilde{f}_x$ to f_x, accepting arguments from the $\mathbb{R}^s$ in the last line. Let us start with $f_x(\zeta) = c^\top x + \Phi(\zeta - Tx)$, $\forall \zeta \in \mathbb{R}^s$. By Theorem 1.1, (A0)–(A2) ensure that the value function Φ is real-valued and lower semicontinuous. For finiteness of the integrands in (2.4) we assume

$$\text{(A3)} \qquad \int_{\mathbb{R}^s} \|\zeta\| \, \mu(d\zeta) < \infty, \; \int_{\mathbb{R}} |\alpha| \, \nu(d\alpha) < \infty \qquad \text{(finite first moments)}$$

(A2) implies $\Phi(0) = 0$, because 0 is feasible and assuming $\Phi(0) < 0$ yields that q has a negative component with a non-zero coefficient that is not eliminated by other summands. Linear scaling of y then results in Φ's unboundedness, which is a contradiction to (A2). Theorem 1.1, (4) yields for fixed x, that there exist constants $\sigma, \kappa > 0$ such that $\forall \, \zeta \in \mathbb{R}^s$

$$|\Phi(\zeta - Tx) - \Phi(0)| \leq \alpha \|\zeta - Tx\| + \beta$$
$$\Rightarrow \quad |\Phi(\zeta - Tx)| \leq \alpha \|\zeta\| + \sigma$$
$$\Rightarrow \quad \left| f_x(\zeta) - c^\top x \right| \leq \alpha \|\zeta\| + \sigma$$
$$\Rightarrow \quad \left| [f_x(\zeta) - c^\top x]_+ \right| \leq \alpha \|\zeta\| + \sigma$$
$$\Rightarrow \quad |[f_x(\zeta) - \eta]_+| \leq \alpha \|\zeta\| + \kappa.$$

Hence (A0)–(A3) imply that the integral on the left in (2.4) is always finite. For the integral on the right (A3) ensures this property. In accordance with (A3) we denote by $\mathscr{P}_1(\mathbb{R}^s), \mathscr{P}_1(\mathbb{R})$ the subsets of $\mathscr{P}(\mathbb{R}^s), \mathscr{P}(\mathbb{R})$ with measures having finite first moments.

We now fix $\nu \in \mathscr{P}_1(\mathbb{R})$ and consider the *multifunction* (set-valued mapping)

$$\begin{aligned} C: \quad & \mathscr{P}_1(\mathbb{R}^s) \longrightarrow 2^{\mathbb{R}^m} \\ & \mu \longmapsto \{x \in \mathbb{R}^m \; : \; \tilde{f}_x \leq_{icx} a, \, x \in X\}. \end{aligned} \qquad (2.5)$$

It might look a bit odd to the reader, that μ is not appearing in the definition of its image. However the impact of μ is hidden in $\tilde{f}_x$. A variation of μ coincides with a variation of z which controls $\tilde{f}_x$ and hence the set to which μ is mapped.

Denote by $C(S)$ the set of all bounded and continuous real functions defined on some normal topological space S. By the next definition the space $\mathscr{P}_1(\mathbb{R}^s)$ is equipped with weak convergence of probability measures ([17]).

Definition 2.1. A sequence $(\mu_n)_{n=1}^{\infty}$ in $\mathscr{P}_1(\mathbb{R}^s)$ is said to *converge weakly* to $\mu \in \mathscr{P}_1(\mathbb{R}^s)$, written $\mu_n \xrightarrow{w} \mu$, iff for any (test) function $h \in C(\mathbb{R}^s)$ it holds $\int_{\mathbb{R}^s} h(z)\,\mu_n(dz) \longrightarrow \int_{\mathbb{R}^s} h(z)\,\mu(dz)$ as $n \to \infty$.

Remark 2.2. A norm in $C(S)$ is given by $|f| := \sup_{s \in S} |f(s)|$. The proof of Theorem IV 6.2.2 in [41] shows that there is an isometric isomorphism between $C(S)'$ and the linear space $rba(S)$ of all regular bounded additive set functions defined on the field generated by the closed subsets of S such that corresponding elements x' and μ satisfy the identity $x'f = \int_S f(s)\,\mu(ds) \ \forall f \in C(S)$. In other words the dual (or conjugate) space $C(S)'$ of $C(S)$ can be identified with $rba(S)$. In [41, IV.15] a list of dual spaces can be found. A sequence $(x_k')_{k \in \mathbb{N}}$ in $C(S)'$ is said to *converge weakly** to x' in $C(S)'$, iff $x_k'f \longrightarrow x'f \ \forall f \in C(S)$. Hence weak convergence of probability measures in $\mathscr{P}_1(\mathbb{R}^s)$ corresponds to weak*-convergence in $C(\mathbb{R}^s)' \cong rba(\mathbb{R}^s) \supseteq \mathscr{P}_1(\mathbb{R}^s)$.

Remark 2.3. Due to the fact that the integrals $\int h\,d\mu$ completely determines μ (see [17], Theorem 1.3) weak limits are unique.

The next theorem provides equivalent conditions to weak convergence of probability measures. In our context condition (iv) will be especially useful.

Theorem 2.4. *(Portmanteau Theorem (see [17], Theorem 2.1, p. 11/12)) Let S be a metric space and let μ_n, μ be probability measures on $(S, \mathscr{B}(S))$. These five conditions are equivalent:*

> *(i)* $\mu_n \xrightarrow{w} \mu$.
> *(ii)* $\lim_n \int h\,d\mu_n = \int h\,d\mu$ *for all bounded, uniformly continuous real h.*
> *(iii)* $\limsup_n \mu_n(F) \leq \mu(F)$ *for all closed F.*
> *(iv)* $\liminf_n \mu_n(G) \geq \mu(G)$ *for all open G.*
> *(v)* $\lim_n \mu_n(A) = \mu(A)$ *for all $A \in \mathscr{B}(S)$ with $\mu(\partial A) = 0$ (μ-continuity sets).* *Note that the boundary of A, ∂A is closed.*

In the case $S = \mathbb{R}$ with its usual topology, if μ_n, μ denote the probability measures generated by arbitrary distribution functions F_n, F respectively, then (i)–(v) are also equivalent to convergence in distribution (see [17, p. 2])

> *(vi)* $\lim_n F_n(x) = F(x)$ *for all points $x \in \mathbb{R}$ at which F is continuous.*

Example 2.5. Choosing real random variables X_n, Y_n according to $\mathbb{P}(X_n = 1) = 1$ for all $n \in \mathbb{N}$ and $\mathbb{P}(Y_n = n) = 1/n$ and $\mathbb{P}(Y_n = 0) = 1 - 1/n$ shows that $\leq_{icx}$ is not *closed with respect to weak convergence*, i.e., $X_n \leq_{icx} Y_n \ \forall n \in \mathbb{N}$, but $\lim_n X_n \not\leq_{icx} \lim_n Y_n$. The reason for that is, that Y_n indeed converges in distribution but $\mathbb{E}(Y_n) \not\to \mathbb{E}(Y)$ (cf. [79, Example 1.5.8]).

Our aim now is to show that the optimal value function $\varphi : \mathscr{P}_1(\mathbb{R}^s) \ni \mu \mapsto \inf\{g(x) : x \in C(\mu)\} \in \mathbb{R} \cup \{\pm\infty\}$ is lower semicontinuous for which the key is to prove that C is a closed multifunction on $\mathscr{P}_1(\mathbb{R}^s)$. Closedness of C is a continuity concept for multifunctions, which roughly spoken means that limit points belong to limit sets. As a consequence $C(\mu_n)$ does not abruptly collapse back upon itself while passing over to the limit set.

Definition 2.6. Let (X, d_X) and (Λ, d_λ) be metric spaces.[14] A point-to-set mapping (multivalued mapping, multifunction) $\Gamma : \Lambda \to 2^X$ is *closed* at a point $\lambda_0 \in \Lambda$, iff for each sequence $(\lambda_t, x_t)_{t=1}^\infty$ in $\Lambda \times X$ with the properties

$$\lambda_t \to \lambda_0 , \quad x_t \in \Gamma(\lambda_t) , \quad x_t \to x_0$$

it follows that $x_0 \in \Gamma(\lambda_0)$.

The following lemma enables us to prove the closedness of C in a concise way. Due to the similarity to the implication (i)$\Rightarrow$(iv) from Theorem 2.4, we call it "Portmanteau Theorem for integrals".

Lemma 2.7. *("Portmanteau Theorem for integrals")*
Let $\mu_n, \mu \in \mathscr{P}(\mathbb{R}^s)$ *with* $\mu_n \overset{w}{\longrightarrow} \mu$ *and* $h : \mathbb{R}^s \to \mathbb{R}$ *be lower semicontinuous with* $h(z) \geq 0 \ \forall z \in \mathbb{R}^s$. *Then*

$$\int_{\mathbb{R}^s} h(z)\, \mu(dz) \leq \liminf_n \int_{\mathbb{R}^s} h(z)\, \mu_n(dz).$$

Proof. We start with the bounded case and assume there exist $\underline{h}, \overline{h} \in \mathbb{R}$ such that $\underline{h} < h(z) < \overline{h} \ \forall z \in \mathbb{R}^s$. Without loss of generality we assume $0 < h(z) < 1 \ \forall z \in \mathbb{R}^s$ which can be achieved by affine scaling according to $t \mapsto (t - \underline{h})/(\overline{h} - \underline{h})$. Fix $k \in \mathbb{N}$ and consider the sets $H_i := \{z \in \mathbb{R}^s : i/k < h(z)\}, i = 0, \ldots, k$. Since h is lower semicontinuous, H_i is open (and measurable) for all i.

It holds

$$\sum_{i=1}^k \frac{i-1}{k}\, \mu\left(H_{i-1} \cap \overline{H}_i\right) \leq \int_{\mathbb{R}^s} h(z)\, \mu(dz) \leq \sum_{i=1}^k \frac{i}{k}\, \mu\left(H_{i-1} \cap \overline{H}_i\right),$$

---

[14]For our purposes a notion of convergence on certain spaces of probability measures is already sufficient. That is X and Λ could be considered as topological spaces. If we topologize those spaces of probability measures by taking as the general basic neighborhood of P the set of Q such that $|\int f_i dP - \int f_i dQ| < \varepsilon$ for $i = 1, \ldots, k$, where ε is positive and the f_i lie in $C(S)$, then weak convergence is convergence in this topology (see [17, p. 11]). On the other hand the spaces of interest with this topology are also metrizable.

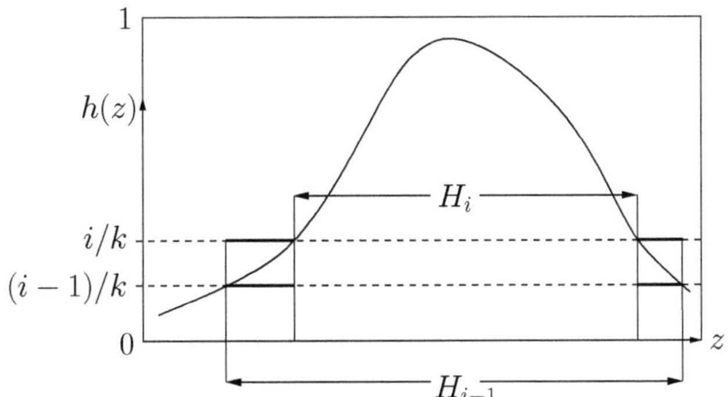

Figure 2.1: Approximation of the μ-integral of h as in the proof of Lemma 2.7

where $\overline{H}_i$ denotes the set-theoretic complement of H_i in $\mathbb{R}^s$. Since $H_i \subset H_{i-1}$ $(*)$ and $\mu(H_0) = 1$ and $\mu(H_k) = 0$ $(**)$, the sum on the right equals

$$\sum_{i=1}^{k} \frac{i}{k}\, \mu(H_{i-1} \setminus H_i) \overset{(*)}{=} \sum_{i=1}^{k} \frac{i}{k}\, \left(\mu(H_{i-1}) - \mu(H_i)\right) \overset{(**)}{=} \frac{1}{k} + \frac{1}{k}\sum_{i=1}^{k} \mu(H_i),$$

while the sum on the left is identical with

$$\sum_{i=1}^{k} \frac{i-1}{k}\, \left(\mu(H_{i-1}) - \mu(H_i)\right) \;=\; \frac{1}{k}\sum_{i=1}^{k} \mu(H_i).$$

Putting this together yields

$$\frac{1}{k}\sum_{i=1}^{k} \mu(H_i) \;\leq\; \int_{\mathbb{R}^s} h(z)\, \mu(dz) \;\leq\; \frac{1}{k} + \frac{1}{k}\sum_{i=1}^{k} \mu(H_i). \tag{2.6}$$

By the Portmanteau Theorem (2.4, (iv)) we have for all i

$$\mu(H_i) \;\leq\; \liminf_{n} \mu_n(H_i). \tag{2.7}$$

Applying the left inequality in (2.6) to μ_n and taking the limes inferior provides

$$\frac{1}{k}\liminf_{n} \sum_{i=1}^{k} \mu_n(H_i) \;\leq\; \liminf_{n} \int_{\mathbb{R}^s} h(z)\, \mu_n(dz),$$

and, together with (2.7),

$$\frac{1}{k}\sum_{i=1}^{k} \mu(H_i) \;\leq\; \liminf_{n} \int_{\mathbb{R}^s} h(z)\, \mu_n(dz).$$

Now we apply the right inequality in (2.6) and obtain

$$-\frac{1}{k} + \int_{\mathbb{R}^s} h(z)\,\mu(dz) \;\leq\; \liminf_n \int_{\mathbb{R}^s} h(z)\,\mu_n(dz).$$

With $k \to \infty$ this yields the assertion for bounded h. For extension to unbounded non-negative h let $r \in \mathbb{R}_+$ and consider the truncated function $h_r : \mathbb{R}^s \to \mathbb{R}$ with

$$h_r(\,.\,) := \begin{cases} h(\,.\,)\,, & \text{if } h(\,.\,) \leq r \\ r\,, & \text{otherwise.} \end{cases}$$

Lower semicontinuity of h implies lower semicontinuity of h_r for all $r \in \mathbb{R}_+$. The assertion then is valid for h_r, because h_r is bounded. Moreover, $h_r(z) \leq h(z)\,\forall z \in \mathbb{R}^s$. This yields

$$\begin{aligned} &\int_{\mathbb{R}^s} h_r(z)\,\mu(dz) \\ \leq\;& \liminf_n \int_{\mathbb{R}^s} h_r(z)\,\mu_n(dz) && (2.8) \\ \leq\;& \liminf_n \int_{\mathbb{R}^s} h(z)\,\mu_n(dz) && \forall r \in \mathbb{R}_+. \end{aligned}$$

The Monotone Convergence Theorem (see for instance [18], Theorem 16.2, p. 211) yields

$$\int_{\mathbb{R}^s} h_r(z)\,\mu(dz) \;\longrightarrow\; \int_{\mathbb{R}^s} h(z)\,\mu(dz) \qquad \text{for } r \to \infty.$$

Together with (2.8) this implies

$$\int_{\mathbb{R}^s} h(z)\,\mu(dz) \;\leq\; \liminf_n \int_{\mathbb{R}^s} h(z)\,\mu_n(dz),$$

and the proof is complete. $\qquad\square$

Proposition 2.8. *Assume (A0)–(A3). Then the multifunction C, as defined in (2.5), is closed on $\mathscr{P}_1(\mathbb{R}^s)$.*

Proof. Let $\mu_n, \mu \in \mathscr{P}_1(\mathbb{R}^s)$ and $x_n \in C(\mu_n)$ such that $\mu_n \xrightarrow{\;w\;} \mu$ and $x_n \to x$. Closedness of X then immediately yields $x \in X$. According to (2.4), $x_n \in C(\mu_n)$ implies

$$\int_{\mathbb{R}^s} [f_{x_n}(\zeta) - \eta]_+\,\mu_n(d\zeta) \;\leq\; \int_{\mathbb{R}} [a - \eta]_+\,v(da) \qquad \forall \eta \in \mathbb{R}. \qquad (2.9)$$

Notice that the integrands on the left are non-negative and lower semicontinuous with respect to x_n and ζ for all $\eta \in \mathbb{R}$. Together with Fatou's Lemma (see for instance [18], Theorem 16.3, p. 212), this implies

$$
\begin{aligned}
\int_{\mathbb{R}^s} [f_x(\zeta) - \eta]_+ \, \mu_n(d\zeta) &\leq \int_{\mathbb{R}^s} \liminf_k [f_{x_k}(\zeta) - \eta]_+ \, \mu_n(d\zeta) \\
&\leq \liminf_k \int_{\mathbb{R}^s} [f_{x_k}(\zeta) - \eta]_+ \, \mu_n(d\zeta)
\end{aligned}
$$

for all $\eta \in \mathbb{R}$. Taking the limes inferior with respect to n on both sides we obtain

$$
\begin{aligned}
\liminf_n \int_{\mathbb{R}^s} [f_x(\zeta) - \eta]_+ \, \mu_n(d\zeta) &\leq \liminf_n \liminf_k \int_{\mathbb{R}^s} [f_{x_k}(\zeta) - \eta]_+ \, \mu_n(d\zeta) \\
&\leq \liminf_n \int_{\mathbb{R}^s} [f_{x_n}(\zeta) - \eta]_+ \, \mu_n(d\zeta) \\
&\leq \int_{\mathbb{R}} [a - \eta]_+ \, \nu(da) \qquad \forall \eta \in \mathbb{R}.
\end{aligned}
$$

Here the second inequality follows from passing to a subsequence of (x_k, μ_n), namely the diagonal sequence where $n = k$, and the third inequality follows from (2.9). Applying the Portmanteau-type Lemma 2.7 with $h(\zeta) := [f_x(\zeta) - \eta]_+$ implies $\forall \eta \in \mathbb{R}$

$$
\begin{aligned}
\int_{\mathbb{R}^s} [f_x(\zeta) - \eta]_+ \, \mu(dz) & \\
\leq \liminf_n \int [f_{x_n}(\zeta) - \eta]_+ \, \mu_n(dz) & \\
\leq \int_{\mathbb{R}} [a - \eta]_+ \, \nu(da) &
\end{aligned}
$$

and thus according to (2.4), $x \in C(\mu)$, what completes the proof. $\qquad\square$

Remark 2.9. (About closedness of the sets $C(\mu)$.) By setting μ_n identical to μ for all n, Proposition 2.8 implies that $C(\mu)$ is a closed subset of $\mathbb{R}^m$ for all $\mu \in \mathscr{P}_1(\mathbb{R}^s)$.

Remark 2.10. (About convexity of the sets $C(\mu)$.) Assume that X is convex and that there are no integer variables in the second stage, i.e., $\Phi(t) = \min\{q^\top y : Wy = t, y \geq 0\}$ $\forall t$. The convexity of Φ (*), recall (1.11), then implies that for all $x_1, x_2 \in X$ and all λ with $0 \leq \lambda \leq 1$

$$
[f_{\lambda x_1 + (1-\lambda)x_2}(\zeta) - \eta]_+ \overset{(*)}{\leq} [\lambda(f_{x_1}(\zeta) - \eta) + (1-\lambda)(f_{x_2}(\zeta) - \eta)]_+
$$

$$\leq \quad \lambda\left[f_{x_1}(\zeta) - \eta\right]_+ + (1-\lambda)\left[f_{x_2}(\zeta) - \eta\right]_+.$$

The second "$\leq$" follows from case differentiation. Integration, together with (2.4) yields the convexity of $C(\mu)$ for all $\mu \in \mathscr{P}_1(\mathbb{R}^s)$.

Remark 2.11. (About variable v.) In [35] the authors have studied the stability of first order stochastic dominance constraints (involving generic random variables) when perturbing the underlying probability distributions for the data *and* the benchmark. When equipping the space $\mathscr{P}_1(\mathbb{R})$ of benchmark measures v with weak convergence of probability measures and selecting the benchmarks from the subset $\mathscr{P}_{\rho,R}(\mathbb{R}) \subset \mathscr{P}_1(\mathbb{R})$ of measures whose ρ-th moment is bounded from above by R ($\rho > 1, R > 0$ fixed), then $v_n, v \in \mathscr{P}_{\rho,R}(\mathbb{R})$ and $v_n \xrightarrow{w} v$ imply $\int_{\mathbb{R}}[a - \eta]_+ v_n(da) \to \int_{\mathbb{R}}[a - \eta]_+ v(da)$:

$$v_n \xrightarrow{w} v$$
$$\Leftrightarrow \quad \mathbb{P} \circ a_n^{-1} \xrightarrow{w} \mathbb{P} \circ a^{-1}$$
$$\Leftrightarrow \quad a_n \xrightarrow{\mathscr{D}} a \quad \text{(Convergence in Distribution, cf. [17])}$$
$$\Rightarrow \quad [a_n - \eta]_+ \xrightarrow{\mathscr{D}} [a - \eta]_+.$$

Let $\rho := 1 + \varepsilon$

$$\int_{\{a_n \geq \gamma\}} |a_n| \, d\mathbb{P} \leq \frac{1}{\gamma^\varepsilon} \int_{\{a_n \geq \gamma\}} |a_n|^{1+\varepsilon} \, d\mathbb{P} \leq \frac{1}{\gamma^\varepsilon} \mathbb{E}(|a_n|^{1+\varepsilon}) \leq \frac{R}{\gamma^\varepsilon} \quad \forall n \in \mathbb{N}$$
$$\Rightarrow \quad \limsup_{\gamma \to \infty} \int_{\{|a_n| \geq \gamma\}} |a_n| \, d\mathbb{P} = 0,$$

i. e., the a_n and $[a_n - \eta]_+$ are *uniformly integrable*. By Theorem 5.4, p. 32 in [17] this implies the assertion. This enables straightforward extension of the proof of Proposition 2.8 to the multifunction $\bar{C} : \mathscr{P}_1(\mathbb{R}^s) \times \mathscr{P}_{\rho,R}(\mathbb{R}) \to 2^{\mathbb{R}^m}$ where $\bar{C}(\mu, v)$ is defined as $\{x \in \mathbb{R}^m : f_x \leq_{icx} a, x \in X\}$. Only the premises have to be adapted.

As already mentioned, closedness of the multifunction C is the key to proving lower semicontinuity of the optimal value function given by

$$\varphi(\mu) := \inf\{g(x) : x \in C(\mu)\} \quad \forall \mu \in \mathscr{P}_1(\mathbb{R}^s).$$

Proposition 2.12. (*About lower semicontinuity of the optimal value.*) *Assume* (A0)–(A3), *that* $\emptyset \neq X$ *is compact and that* g *is lower semicontinuous. Let* $\bar{\mu} \in \mathscr{P}_1(\mathbb{R}^s)$ *be such that* $\min\{g(x) : x \in C(\bar{\mu})\}$ *has an optimal solution. Then the optimal value function* $\varphi(\mu) := \inf\{g(x) : x \in C(\mu)\}$ *is lower semicontinuous at* $\bar{\mu}$ *(cf. [10] or [52]).*

Proof. Let $\mu_n \xrightarrow{w} \bar{\mu}$ and assume without loss of generality that $C(\mu_n) \neq \emptyset$ for all n. Otherwise, we would have $\varphi(\mu_n) = +\infty$ which does not interfere with the validity of $\liminf_n \varphi(\mu_n) \geq \varphi(\bar{\mu})$.

Let $\varepsilon > 0$ be arbitrarily fixed. Then there exist $x_n \in C(\mu_n)$ such that $g(x_n) \leq \varphi(\mu_n) + \varepsilon$. By compactness of X there exists an accumulation point $\bar{x} \in X$ of the x_n. By closedness of C (Proposition 2.8), it follows that $\bar{x} \in C(\bar{\mu})$. Together with g's lower semicontinuity this implies

$$\varphi(\bar{\mu}) \overset{\bar{x}\in C(\bar{\mu})}{\leq} g(\bar{x}) \overset{(g\ \mathrm{lsc})}{\leq} \liminf_n g(x_n) \leq \liminf_n \varphi(\mu_n) + \varepsilon.$$

Since $\varepsilon > 0$ was arbitrary, the proof is complete. $\qquad\qquad\qquad\square$

At the beginning of this chapter we mentioned that our main object of investigation (2.1) is related to the model

$$\min_{x\in X} \left\{ g(x) : \tilde{f}_x \leq_{st} a \right\}, \tag{2.10}$$

where "$\leq_{icx}$" is replaced by the usual stochastic order. In [52], results similar to those we presented here were published. In particular closedness of the constraint set mapping with the underlying probability measure as parameter and consequently the lower semicontinuity of the optimal value function could be established. In the case of finite probability spaces, (2.10) was shown to be equivalent to a large-scale, block-structured, mixed-integer linear program. This is what we are going to do next for (2.1). Furthermore a decomposition algorithm for these structured mixed-integer linear programming equivalents was proposed.

In [40] (see also [69]) it is assumed that the optimization problem behind Φ (see (1.5)) is a *linear program*. The authors propose a cutting-plane algorithm for (2.10) in the linear recourse setting which employs the ideas of the traditional L-shaped method for stochastic programs ([20, 65, 94, 100]). The original problem is approached by tighter and tighter relaxations, and out of a huge variety of cuts only those are generated, that are needed for the progress of the method. This enables a shortcut over the more generally valid algorithm in [52] and over application of general-purpose MIP solvers.

2.2 Deterministic Equivalents

For discrete probability distributions, the following proposition establishes an equivalence between (2.1) and a large-scale mixed-integer linear program, which, with a view to its algorithmic treatment, features an advantageous structure.

Proposition 2.13. *Let z and a in (2.1) follow discrete distributions with only finitely many realizations z_ℓ, $\ell = 1,\ldots,L$, and a_k, $k = 1,\ldots,K$, as well as probabilities π_ℓ, $\ell = 1,\ldots,L$, and p_k, $k = 1,\ldots,K$, respectively. Let further g be linear. Assume (A1) and (A2). Then (2.1) is equivalent to the mixed-integer linear program*

$$\min\left\{ g^\top x : \begin{array}{llll} c^\top x + q^\top y_{\ell k} - v_{\ell k} & \leq a_k & \forall \ell & \forall k \\ Tx + Wy_{\ell k} & = z_\ell & \forall \ell & \forall k \\ \sum_{\ell=1}^{L} \pi_\ell v_{\ell k} & \leq \mathbb{E}[(a - a_k)_+] & & \forall k \\ x \in X,\ y_{\ell k} \in Y,\ & v_{\ell k} \geq 0 & \forall \ell & \forall k \end{array} \right\} \qquad (2.11)$$

To prove Proposition 2.13, we first show that, for $\tilde{f}_x \leq_{icx} a$, validity of the equivalent relation $\int_{\mathbb{R}^s}[f_x(\zeta) - \eta]_+ \mu(d\zeta) \leq \int_{\mathbb{R}}[a - \eta]_+ v(da)\ \forall \eta \in \mathbb{R}$ is already sufficient if it holds at the breakpoints $a_1,\ldots,a_K$ of $F_a^{(2)}(\eta)$ with

$$F_a^{(2)}(\eta) := \mathbb{E}((a - \eta)_+) = \int_\eta^\infty [a(\omega) - \eta]_+\, \mathbb{P}(d\omega) = \int_{\mathbb{R}} [a - \eta]_+\, v(da)\ \forall \eta \in \mathbb{R}.$$

For this purpose, we need the following lemma:

Lemma 2.14. *Let X be a discretely distributed random variable with finitely many realizations. Then $F_X^{(2)}$ is a piecewise linear Lipschitzian function, which is differentiable in all points where F_X is continuous. Furthermore, $F_X^{(2)}$ is monotonously non-increasing and convex.*

Proof. W.l.g. let $y_1 < \ldots < y_m$ be the realizations of X, $\eta \in \mathbb{R}$ and

$$j := \min\{\{i \in \{1,\ldots,m\} : \eta < y_i\} \cup \{m\}\}.$$

Furthermore let $y_0 < y_1$. Then

$$\begin{aligned} F_X^{(2)}(\eta) &= \int_\eta^\infty 1 - F_X(z)\, dz \\ &= \int_\eta^{y_m} 1 - F_X(z)\, dz \\ &= \int_\eta^{y_j} 1 - F_X(z)\, dz + \int_{y_j}^{y_m} 1 - F_X(z)\, dz. \end{aligned}$$

The integrands are step functions, which can be defined with respect to the subdivisions $t < y_j$ and $y_j < y_{j+1} < \ldots < y_m$ respectively. With $p_i := \mathbb{P}(y_i)$ integration yields $\forall t \in \mathbb{R}$

$$F_X^{(2)}(t) = \left(1 - \sum_{i=1}^{j-1} p_i\right)(y_j - t) + \sum_{i=j}^{m-1}\left(1 - \sum_{k=1}^{i} p_k\right)(y_{i+1} - y_i),$$

which obviously is affine on each $\left[y_{j-1}, y_j\right]$ and hence continuous on $\mathbb{R}$. Its derivative in $\left(y_{j-1}, y_j\right)$ is $\sum_{i=1}^{j-1} p_i - 1 \in [-1, 0]$. $\qquad\square$

As announced, the next result shows that it is possible to significantly thin out the continuum of constraints in (2.4) under the prerequisite of Lemma 2.14 applying to the benchmark profile:

Corollary 2.15. *Let* X, Y *be real random variables on some probability space and let* Y *have only finitely many realizations* $y_1 < \ldots < y_K$. *Then it holds*

$$X \leq_{icx} Y$$
$$\Longleftrightarrow \quad \mathbb{E}\left[(X - y_k)_+\right] \leq \mathbb{E}\left[(Y - y_k)_+\right], \quad k = 1, \ldots, K.$$

This is a consequence of the convexity of the integrated survival functions (as functions of the lower integration bounds) and of its piecewise linearity in the discrete case. The linearity regions can be seen as shortened secants (cf. Figure 2.3)

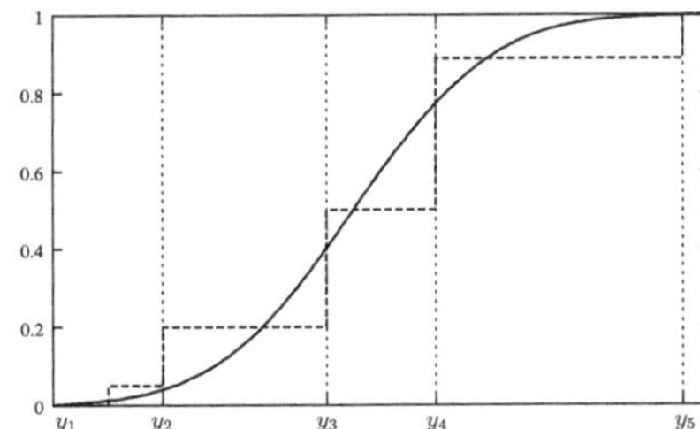

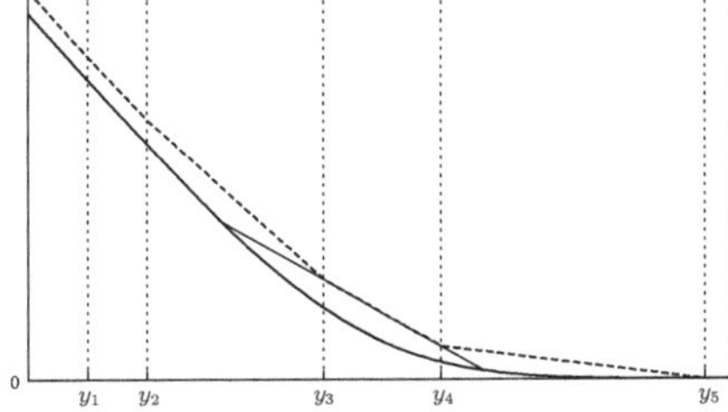

Figure 2.2: $X \leq_{icx} Y$ with Y only taking on the values $y_1, \ldots, y_5$.

Figure 2.3: It is sufficient to check the relation $F_X^{(2)}(t) \leq F_Y^{(2)}(t)$ for $t \in \{y_1, \ldots, y_5\}$.

Proof. Of course "$\Rightarrow$" follows immediately from the definition of $\leq_{icx}$. For the other implication let us consider three cases depending on the values of η.

Case 1. $\eta < y_1$

$$F_Y^{(2)}(\eta) = -\int_\eta^{y_K} P(Y \leq z)\, dz + y_K - \eta$$

$$
\begin{aligned}
&= && -\int_{y_1}^{y_K} P(Y \le z)\, dz + y_K - \eta \\
&= && F_Y^{(2)}(y_1) - \eta + y_1 \\
&\overset{\text{(As.)}}{\ge} && F_X^{(2)}(y_1) - \eta + y_1 \\
&= && \int_{y_1}^{\infty} 1 - P(X \le z)\, dz - \eta + y_1 \\
&\ge && -\int_{\eta}^{y_1} P(X \le z)\, dz + y_1 - \eta + \int_{y_1}^{\infty} 1 - P(X \le z)\, dz \\
&= && \int_{\eta}^{\infty} 1 - P(X \le z)\, dz \\
&= && F_X^{(2)}(\eta)
\end{aligned}
$$

Case 2. $\eta \in [y_i, y_{i+1}]$, $1 \le i \le k-1$, $\lambda := \frac{y_{i+1}-\eta}{y_{i+1}-y_i}$

$$
\begin{aligned}
F_X^{(2)}(\eta) \quad &\overset{\text{(Conv.)}}{\le} && \lambda F_X^{(2)}(y_i) + (1-\lambda)F_X^{(2)}(y_{i+1}) \\
&\overset{\text{(As.)}}{\le} && \lambda F_Y^{(2)}(y_i) + (1-\lambda)F_Y^{(2)}(y_{i+1}) \\
&\overset{\text{(Lin.)}}{=} && F_Y^{(2)}(\eta)
\end{aligned}
$$

Case 3. $\eta > y_K$

$$
F_X^{(2)}(\eta) \overset{\text{(Mon.)}}{\le} F_X^{(2)}(y_K) \overset{\text{(As.)}}{\le} F_Y^{(2)}(y_K) = F_Y^{(2)}(\eta) \qquad\qquad \square
$$

Proof of Proposition 2.13: To establish the asserted equivalence we fix k, consider the sets

$$
S_1 := \left\{ x \in X : \int_{\mathbb{R}^s} [f_x(\zeta) - a_k]_+\, \mu(d\zeta) \le \int_{\mathbb{R}} [a - a_k]_+\, v(da) \right\}
$$

and

$$
S_2 := \left\{ x \in X : \begin{aligned} &\exists v_\ell \ge 0,\ \exists y_\ell \in Y,\quad \ell = 1,\dots,L, \\ &\text{such that:} \\ &\quad c^\top x + q^\top y_\ell - v_\ell && \le a_k \\ &\quad Tx + Wy_\ell && = z_\ell \\ &\quad \textstyle\sum_{\ell=1}^{L} \pi_\ell v_\ell && \le \mathbb{E}[(a - a_k)_+] \end{aligned} \right\}
$$

and show that $S_1 = S_2$.

For $S_1 \subset S_2$ let $x \in S_1$ and denote $I := \{\ell \in \{1,\ldots,L\} : f_x(z_\ell) - a_k > 0\}$. By the definition of S_1 we have

$$\int_{\mathbb{R}^s} [f_x(\zeta) - a_k]_+ \, \mu(d\zeta) \;=\; \sum_{\ell \in I} \pi_\ell (f_x(z_\ell) - a_k) \;\leq\; \mathbb{E}[(a - a_k)_+].$$

Put $v_\ell := f_x(z_\ell) - a_k$ for all $l \in I$, and $v_\ell := 0$, otherwise. This yields

$$\sum_{\ell=1}^{L} \pi_\ell v_\ell \;\leq\; \mathbb{E}[(a - a_k)_+].$$

For $\ell \notin I$ it holds that $f_x(z_\ell) - a_k \leq 0$. The validity of (A0)–(A2) implies that the optimization problems behind the $f_x(z_\ell)$ are solvable. Hence, for all $\ell \notin I$, there exist $y_\ell \in Y$ with

$$c^\top x + q^\top y_\ell - a_k \leq 0 = v_\ell \quad \text{and} \quad Tx + Wy_\ell = z_\ell.$$

For $\ell \in I$, choose $y_\ell \in Y$ such that $q^\top y_\ell = \Phi(z_\ell - Tx)$ and $Tx + Wy_\ell = z_\ell$. Then

$$c^\top x + q^\top y_\ell - a_k = f_x(z_\ell) - a_k = v_\ell,$$

yielding $x \in S_2$.

For $S_2 \subset S_1$ let $x \in S_2$ and $(v_\ell)_{\ell=1}^{L}$ be a feasible configuration of the v_ℓ. Consider $I := \{\ell \in \{1,\ldots,L\} : v_\ell > 0\}$. The definition of S_2 implies that for $\ell \notin I$ there exist $y_\ell \in Y$ fulfilling

$$c^\top x + q^\top y_\ell - a_k \leq 0 \quad \text{and} \quad Tx + Wy_\ell = z_\ell.$$

Therefore, $f_x(z_\ell) - a_k \leq 0$ for all $\ell \notin I$. For $\ell \in I$ there exist $y_\ell \in Y$ with

$$c^\top x + q^\top y_\ell - a_k \leq v_\ell \quad \text{and} \quad Tx + Wy_\ell = z_\ell.$$

Thus, $f_x(z_\ell) - a_k \leq v_\ell$ for all $\ell \in I$. Now we obtain

$$\int_{\mathbb{R}^s} [f_x(\zeta) - a_k]_+ \, \mu(d\zeta) \;=\; \sum_{\ell \in I} \pi_\ell [f_x(z_\ell) - a_k]_+ + \sum_{\ell \notin I} \pi_\ell [f_x(z_\ell) - a_k]_+$$

$$\overset{(*)}{\leq} \; \sum_{\ell=1}^{L} \pi_\ell v_\ell \;\overset{(x \in S_2)}{\leq}\; \int_{\mathbb{R}} [a - a_k]_+ \, v(da),$$

$(*)$ holds because $\forall \ell \notin I : [f_x(z_\ell) - a_k]_+ = 0$ and because $\pi_\ell v_\ell \geq 0$, $\ell = 1,\ldots,L$. So $x \in S_1$, and the proof is complete. $\qquad\square$

Concerning the above proof we remark that the fact that, for finite probability spaces, the increasing convex order relation reduces to a finite number of linear inequalities, cf. (2.15), has already been observed in [82]. The setting in [82] refers to a different class of random variables and stochastic dominance with preference of big outcomes, though.

Inspecting (2.11) we observe that the constraints

$$\sum_{\ell=1}^{L} \pi_\ell v_{\ell k} \leq \mathbb{E}[(a - a_k)_+] \qquad \forall k \tag{2.12}$$

are the only ones coupling explicitly second-stage variables, namely $v_{\ell k}$, across different scenarios ℓ. An implicit such coupling, of course, is given by

$$\begin{aligned} c^\top x &+ q^\top y_{\ell k} - v_{\ell k} &=& a_k &\forall \ell \, \forall k, \\ Tx &+ W y_{\ell k} &=& z_\ell &\forall \ell \, \forall k. \end{aligned}$$

One concludes that, without (2.12), problem (2.11) in principle were in L-shaped form (cf. [112] and Figure 2.4), a structure that has given rise to different decomposition algorithms for stochastic programs [20, 24, 65, 94, 99, 100].

Understanding (2.11) as an "expanded" representation of the nonconvex global minimization problem (2.1) we propose a branch-and-bound decomposition algorithm for its solution later in this work.

The following corollary points out, that the more or less inconvenient[15] K-multiplicity of the second-stage variables y_ℓ in (2.11) is actually redundant (cf. Figures 2.4 and 2.5).

Corollary 2.16. *Under the assumptions of Proposition 2.13, (2.1) is equivalent to the mixed-integer linear program*

$$\min \left\{ g^\top x : \begin{array}{llll} c^\top x + q^\top y_\ell - v_{\ell k} & \leq a_k & \forall \ell & \forall k \\ Tx + W y_\ell & = z_\ell & \forall \ell & \\ \sum_{\ell=1}^{L} \pi_\ell v_{\ell k} & \leq \mathbb{E}[(a - a_k)_+] & & \forall k \\ x \in X, \, y_\ell \in Y, \, v_{\ell k} \geq 0 & & \forall \ell & \forall k \end{array} \right\} \tag{2.13}$$

---

[15] Actually, instances where CPLEX [60] tackles the "bloated" formulation quicker than the shrunken are known. For example there is a representative (10 data and 4 benchmark scenarios) of the dispersed generation system described in [52] including increasing convex order constraints, where it takes CPLEX approximately 17 seconds to solve the problem of the form (2.11), while the computation (on standard PC) of the smaller problem of the kind (2.13) aborts without a feasible point after about four and a half hour due to lacking main memory.

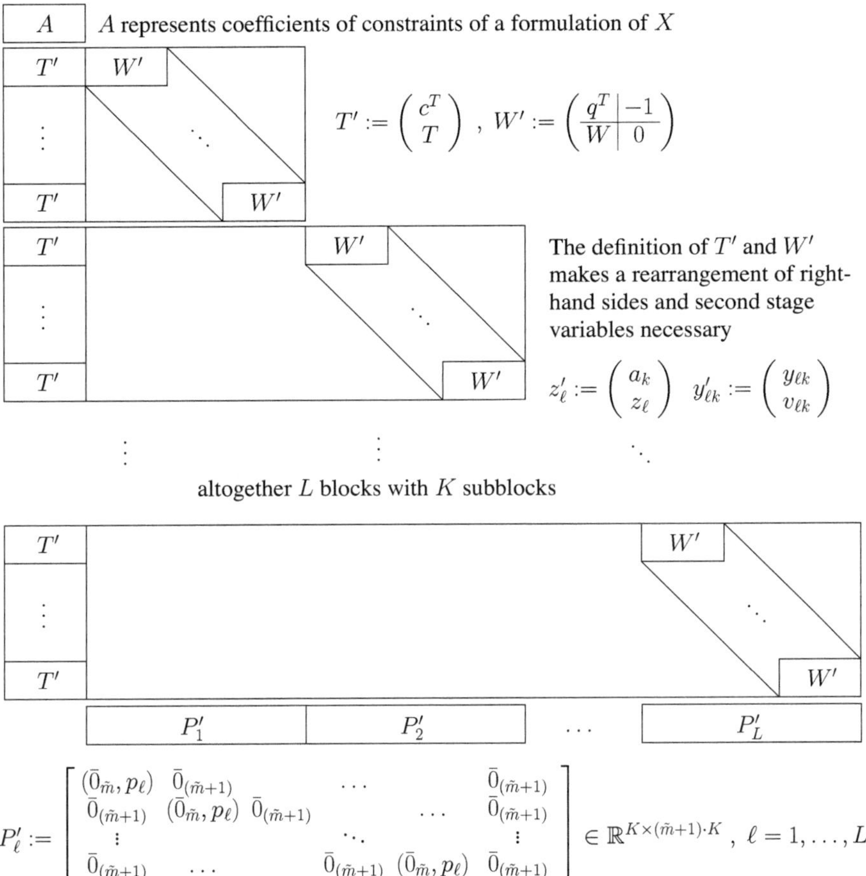

$$T' := \begin{pmatrix} c^T \\ T \end{pmatrix} \ , \ W' := \left(\begin{array}{c|c} q^T & -1 \\ \hline W & 0 \end{array} \right)$$

$$z'_\ell := \begin{pmatrix} a_k \\ z_\ell \end{pmatrix} \quad y'_{\ell k} := \begin{pmatrix} y_{\ell k} \\ v_{\ell k} \end{pmatrix}$$

$$P'_\ell := \begin{bmatrix} (\bar{0}_{\tilde{m}}, p_\ell) & \bar{0}_{(\tilde{m}+1)} & \cdots & & \bar{0}_{(\tilde{m}+1)} \\ \bar{0}_{(\tilde{m}+1)} & (\bar{0}_{\tilde{m}}, p_\ell) & \bar{0}_{(\tilde{m}+1)} & & \cdots & \bar{0}_{(\tilde{m}+1)} \\ \vdots & & \ddots & & & \vdots \\ \bar{0}_{(\tilde{m}+1)} & \cdots & & \bar{0}_{(\tilde{m}+1)} & (\bar{0}_{\tilde{m}}, p_\ell) & \bar{0}_{(\tilde{m}+1)} \\ \bar{0}_{(\tilde{m}+1)} & & \cdots & & \bar{0}_{(\tilde{m}+1)} & (\bar{0}_{\tilde{m}}, p_\ell) \end{bmatrix} \in \mathbb{R}^{K \times (\tilde{m}+1) \cdot K} \ , \ \ell = 1, \ldots, L$$

$\bar{0}_{(\tilde{m}+1)} := (\tilde{m} + 1)$-dimensional (row) nullvector, where $\tilde{m}$ denotes the number of columns of W. $(\bar{0}_{\tilde{m}}, p_\ell) := (\tilde{m} + 1)$-dimensional (row) vector with only the last coordinate different from 0, namely p_ℓ.

Figure 2.4: Structure of the constraint matrix of problem (2.11).

Proof. Let $(x,(y_{\ell k})_{\ell=1,k=1}^{L,K},(v_{\ell k})_{\ell=1,k=1}^{L,K})$ be a feasible point of (2.11). Putting $y_\ell := \arg\min_{k=1,\dots,K}\{q^T y_{\ell k}\}\ \forall \ell$ and keeping x and $(v_{\ell k})_{\ell=1,k=1}^{L,K}$ from (2.11) yields a feasible point of (2.13) with the same objective value.

Conversely, if $(x,(y_\ell)_{\ell=1}^{L},(v_{\ell k})_{\ell=1,k=1}^{L,K})$ is feasible for (2.13) setting $y_{\ell k} := y_\ell$ $\forall \ell,k$, while keeping the other variables is feasible for (2.11). $\qquad\square$

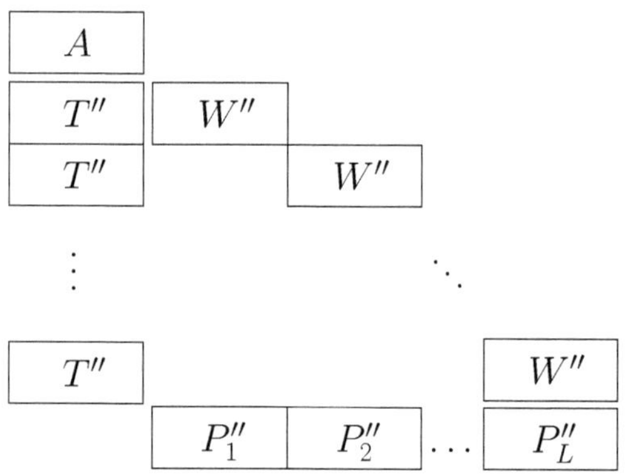

$$T'' := \begin{pmatrix} c^T \\ \vdots \\ c^T \\ T \end{pmatrix}, \quad W'' := \begin{pmatrix} q^T & -1 \\ \vdots & & \ddots \\ q^T & & & -1 \\ \hline W & \bar{0}_s^\top, & \dots, & \bar{0}_s^\top \end{pmatrix}, \quad P_\ell'' := \begin{pmatrix} \bar{0}_{\tilde{m}} & \pi_\ell & 0 & \dots & & 0 \\ & 0 & \pi_\ell & 0 & \dots & 0 \\ \vdots & \vdots & \ddots & \ddots & \ddots & \vdots \\ & 0 & \dots & 0 & \pi_\ell & 0 \\ \bar{0}_{\tilde{m}} & 0 & & \dots & 0 & \pi_\ell \end{pmatrix}$$

$$T'' \in \mathbb{R}^{(K+s)\times m}, \quad W'' \in \mathbb{R}^{(K+s)\times(\tilde{m}+K)}, \quad P_\ell'' \in \mathbb{R}^{K\times(\tilde{m}+K)}$$

Figure 2.5: Structure of the revised constraint matrix in (2.13).

Let us discuss, what is the information content of the solution vector of a model of the type (2.11) or (2.13). First of all the x-part of the solution of these models identifies a member $\tilde{f}_{x^*} \in (\tilde{f}_x)_{x\in X}$ with $g^\top x^*$ being minimal within the subset $\{\tilde{f}_x : \tilde{f}_x \leq_{icx} a,\ x \in X\}$ of the family. This does *not* mean that we have all the information on $\tilde{f}_{x^*}$ in terms of its distribution, since we did not necessarily compute the y-part of $\arg\min\{c^\top x^* + \mathbb{E}(q^\top y_\ell) : \dots\}$. By the solution of the model, we only found a feasible configuration of the y variables yielding a random variable $\hat{f}_{x^*} \leq_{icx} a$ which in general *is not* contained in the family $(\tilde{f}_x)_{x\in X}$ of interest. To also obtain the optimal second-stage policies to the first-stage decision x^* it is necessary to solve $\min_{y_\ell \in Y}\{q^\top y_\ell : Wy_\ell = z_\ell - Tx^*\}$, $\forall \ell$. A simple approach for directly finding the full information on $\tilde{f}_x$ is to modify the objective $g^\top x$ according to $g^\top x + r \cdot \mathbb{E}(q^\top y_\ell)$ with a "small" $r \geq 0$. This would determine an $\tilde{f}_x$ from the family, but not necessarily the one with minimal $g^\top x$—not even for $r \to 0$. Summing up, the scenario strategies should be computed by the solution of the classical (decomposed) expectation-based model with fixed (nonanticipative) first stage x^*.

To complete the section we should specify a deterministic equivalent similar to (2.11) reflecting the preference of larger outcomes instead of smaller ones. In this context (1.2)–(1.5) have to be understood as maximization problems.

Corollary 2.17. *Under the assumptions of Proposition 2.13,*

$$\min\left\{g(x) \ : \ c^\top x + \max_{y \in Y}\left\{q^\top y : Tx + Wy = z(.)\right\} =: \tilde{h}_x(.) \succeq_2 a, \ x \in X\right\}$$

is equivalent to the mixed-integer linear program

$$\min\left\{g^\top x: \begin{array}{llll} c^\top x + q^\top y_\ell + v_{\ell k} & \geq a_k & \forall \ell \ \ \forall k \\ Tx + Wy_\ell & = z_\ell & \forall \ell \\ \sum_{\ell=1}^{L} \pi_\ell v_{\ell k} & \leq \mathbb{E}[(a_k - a)_+] & \forall k \\ x \in X, \ y_\ell \in Y, \ v_{\ell k} \geq 0 & & \forall \ell \ \ \forall k \end{array}\right\} \tag{2.14}$$

The constraints $Tx + Wy_\ell = z_\ell$, $\forall \ell$ and $x \in X$, $y_\ell \in Y$, $\forall \ell$ reflect the feasibility in the definition of $\tilde{h}_x$. Due to the first set of constraints, the variables $v_{\ell k} \geq 0$ now measure the *expected shortfall* of $\tilde{h}_x$ below a_k in scenario ℓ. The constraints containing the weighted sum of the $v_{\ell k}$ variables enforce that the expected shortfall of the profit below a_k does not become too large, namely not larger than the expected shortfall of a below a_k.

To give a formal proof of Corollary 2.17 we first annotate that $X \succeq_2 Y \Leftrightarrow -X \preceq_{icx} -Y$ ($f : \mathbb{R} \to \mathbb{R}$ convex iff $-f$ concave). Furthermore $-\tilde{h}_x(.) = -c^\top x + \min_{y \in Y}\{-q^\top y : Tx + Wy = z(.)\}$. Replacement of the benchmark distribution by the distribution of the negative of a immediately yields (2.14).

Remark 2.18. The quantity $\mathbb{E}[(a - a_k)_+]$ is always non-negative and zero for the largest a_{k^*} among the a_k. Thus the constraint $\sum_{\ell=1}^{L} \pi_\ell v_{\ell k^*} \leq \mathbb{E}[(a - a_{k^*})_+]$ in problem (2.13) implies that $v_{\ell k^*} = 0$ inside the feasible region. Otherwise the convex combination of non-negative values on the left-hand side would be positive. Analogously the related $v_{\ell k}$ in problem (2.14) are equal to zero inside the feasible set. As we will see later, this is algorithmically beneficial.

2.3 Multiple Dominance Constraints

Imagine a group of N decision makers, where each of them has a different acceptance threshold a_i, $i = 1,\dots,N$ for the distribution of $\tilde{f}_x$. A more adequate optimization problem than (2.1) is then given by

$$\min\left\{g^\top x \,:\, \tilde{f}_x \leq_{icx} a_i \,,\, i = 1,\dots,N \,,\, x \in X\right\}. \tag{2.15}$$

A straightforward approach to built up a deterministic equivalent for finitely distributed random outcomes for (2.15) consists of intersecting the feasible sets $\bar{C}(\mu, v_i)$, $i = 1,\dots,N$ each of which is associated to one decision maker:

$$\min\left\{g^\top x:\; \begin{aligned}
c^\top x + q^\top y_\ell - v^i_{\ell k_i} &\leq a^i_{k_i} && \forall \ell \;\; \forall i \;\; \forall k_i \\
Tx + W y_\ell &= z_\ell && \forall \ell \\
\textstyle\sum_{\ell=1}^{L} \pi_\ell v^i_{\ell k_i} &\leq \mathbb{E}[(a - a^i_{k_i})_+] && \forall i \;\; \forall k_i \\
x \in X,\; y_\ell \in Y,\; v^i_{\ell k_i} &\geq 0 && \forall \ell \;\; \forall i \;\; \forall k_i
\end{aligned}\right\} \tag{2.16}$$

Note that the number of benchmark scenarios K depends on the decision maker, i. e., on i. Obviously, the number of variables and constraints in the model might swell a lot. However, things are a bit simpler as in (2.16).

In their article [78] the authors show that the space of probability measures $\mathscr{P}(\mathbb{R})$ (or some subspace) is a *lattice* under most of the known partial orders including first and second order stochastic dominance, the usual stochastic order and the increasing convex order. The point is that if a stochastic order leads to a lattice, then multiple constraints can be equivalently expressed as only one constraint. We first recall the definition of *lattices*.

Definition 2.19. Let $(\mathscr{X}, \succeq_*)$ be an ordered set. For $x, y \in \mathscr{X}$ let $U(x,y) := \{z \in \mathscr{X} \,:\, x \succeq_* z,\, y \succeq_* z\}$ and $V(x,y) := \{z \in \mathscr{X} \,:\, z \succeq_* x,\, z \succeq_* y\}$. If $U(x,y)$ has a smallest element $\tilde{z}$ such that $\tilde{z} \succeq_* z$ for all $z \in U(x,y)$, then $\tilde{z}$ is called the supremum of x and y, denoted by $\tilde{z} = \sup\{x,y\}$. Similarly, if there is a unique largest element z' in $V(x,y)$, then this is called the infimum, denoted by $z' = \inf\{x,y\}$.

If $\sup\{x,y\}$ and $\inf\{x,y\}$ exist for all $x,y \in \mathscr{X}$, then $(\mathscr{X}, \succeq_*)$ is called a lattice. A subset $\mathscr{Y} \subset \mathscr{X}$ of a lattice is called a sublattice if $x, y \in \mathscr{Y}$ implies $\sup\{x,y\} \in \mathscr{Y}$ and $\inf\{x,y\} \in \mathscr{Y}$. Notice that $(\mathscr{Y}, \succeq_*)$ can be a lattice in its own right without being a sublattice.

Remark 2.20. Comparability of orders translates to comparability of suprema and infima in the following way. If $(\mathscr{X}, \succeq_*)$ and $(\mathscr{X}, \succeq_\circledast)$ are lattices and $x \succeq_* y \Rightarrow x \succeq_\circledast y\ \forall x,y \in \mathscr{X}$, then $\inf_*\{x,y\} \succeq_\circledast \inf_\circledast\{x,y\}$ and $\sup_\circledast\{x,y\} \succeq_\circledast \sup_*\{x,y\}$.

For properties of lattices we refer the reader to [4] and [33]. We introduced $\leq_{icx}$ as a relation between real random variables. Notice that $\leq_{icx}$ can also be understood as a partial order on the set of all Borel probability measures by the one-to-one correspondence $\int f \circ X \, d\mathbb{P} \leq \int f \circ Y \, d\mathbb{P} \Leftrightarrow \int f \, d\mathbb{P}_X \leq \int f \, d\mathbb{P}_Y$ already mentioned in footnote 10 on page 8. Another slightly confusing notation in the next theorem concerns the objects in (2.18) and (2.19) which are not contained in $\mathscr{P}_1(\mathbb{R})$. Again a one-to-one relation resolves the inconsistency: $F_X^{(2)}$ yields all the information on the distribution of X:

$$F_X(t_0) = 1 - \lim_{t \searrow t_0} \frac{F_X^{(2)}(t) - F_X^{(2)}(t_0)}{t - t_0}, \ \forall t_0 \in \mathbb{R} \tag{2.17}$$

and hence $\mathbb{P}_X := \mathbb{P} \circ X^{-1}$ is known on $\{]a, b] : a, b \in \mathbb{R}, \ a \leq b\}^{16}$ which is a (intersection-closed) generator of $\mathscr{B}(\mathbb{R})$.

Theorem 2.21. *The ordered set* $(\mathscr{P}_1(\mathbb{R}), \leq_{icx})$ *is a lattice with*

$$F_{\inf\{X,Y\}}^{(2)} = \sup \left\{ g \ \Big| \ \begin{array}{l} g : \mathbb{R} \to \mathbb{R} \text{ is convex and} \\ g(y) \leq \min\{F_X^{(2)}(y), F_Y^{(2)}(y)\}, \ \forall y \in \mathbb{R} \end{array} \right\} \tag{2.18}$$

and

$$F_{\sup\{X,Y\}}^{(2)}(\eta) = \max\{F_X^{(2)}(\eta), F_Y^{(2)}(\eta)\}, \ \forall \eta \in \mathbb{R}. \tag{2.19}$$

$F_{\inf\{X,Y\}}^{(2)}$ is the *convex hull* of the pointwise minimum of $F_X^{(2)}$ and $F_Y^{(2)}$. In other words, $F_{\inf\{X,Y\}}^{(2)}$ is the largest convex function pointwise smaller than or equal to $F_X^{(2)}$ and $F_Y^{(2)}$.

Theorem 2.21 was first shown in [66]. See also [68], [78]. The lattice property of $(\mathscr{P}_1(\mathbb{R}), \leq_{icx})$ is essential for the next proposition, which then follows directly from the definition of an infimum:

Proposition 2.22. *Assume (A0)–(A3), that g is lower semicontinuous and* $X \neq \emptyset$ *is compact. Then it holds*

$$\min\{g(x) : \tilde{f}_x \leq_{icx} a_i, \ i = 1, \ldots, N, \ x \in X\} \tag{2.20}$$

$$= \ \min\{g(x) : \tilde{f}_x \leq_{icx} \inf\{a_i, \ i = 1, \ldots, N\}, \ x \in X\}, \tag{2.21}$$

where the infimum is defined recursively for $N > 2$:

$$\inf\{a_i, \ i = 1, \ldots, N\} := \inf\{\inf\{a_i, \ i = 1, \ldots, N-1\}, a_N\}.$$

---

[16] $\mathbb{P}_X(]a, b]) = F_X(b) - F_X(a)$

Example 2.23. (How a group of decision makers can come to a consensus.) We consider two decision makers in a certain optimization framework. Their different perceptions of what a just acceptable cost distribution could look like is pictured in Figure 2.6. The probabilities of the benchmark realizations correspond to the step heights of the cumulative distribution functions in 2.6 and (after some transformation) to the slopes on the linearity regions of the integrated survival functions in Figure 2.7.

As can be seen from the figure on the right-hand side, the suggested benchmark distributions are not comparable with respect to $\leq_{icx}$, since they intersect each other. The gray graph in Figure 2.7 belongs to the integrated survival function with respect to $F_{\inf\{a_1,a_2\}}$. In Figure 2.6 the gray graph was deduced according to (2.17). It belongs to the cumulative distribution function of the largest random variable Z less than or equal to X and Y in the increasing convex order.

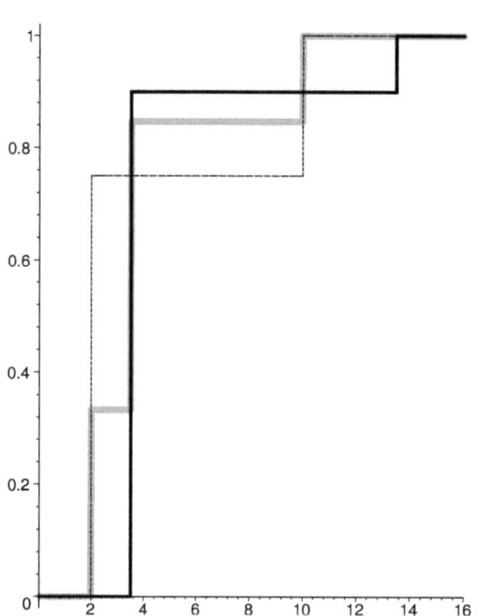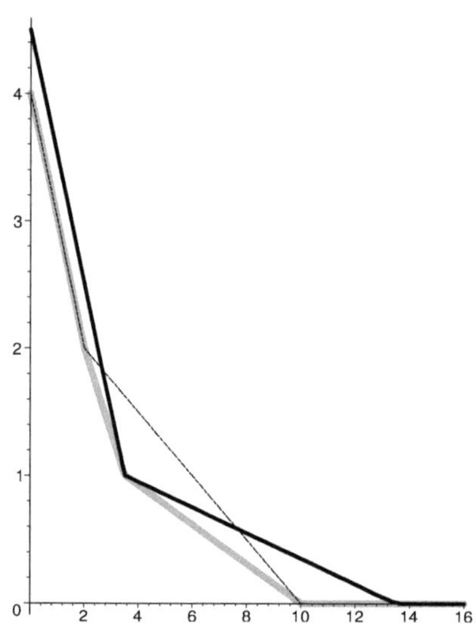

Figure 2.6: Distribution functions proposed by two decision makers. The gray line depicts the distribution function that satisfies both of them.

Figure 2.7: Integrated survival functions $(1 - F_{\inf\{a_1,a_2\}}$ and $1 - F_{a_i}$, $i = 1,2)$ with respect to the 3 cumulative distribution functions in Figure 2.6

Chapter 3

Competitive Risk-Averse Selling Price Determination for Electricity Retailers

3.1 Introduction

To illustrate some computational experience with deterministic equivalents for (2.1) we come to a real-life application of the theory developed so far.

The liberalization of the electricity sector has caused the appearance of a set of electricity markets ([106, 107]). Market agents seek to determine lucrative decisions within these markets to optimize specific objectives. For instance, a generator company desires to optimize the bids to submit to each available electricity market to maximize its selling profit. On the other hand, a large consumer wants to satisfy its demand at minimal electricity procurement costs. A main difficulty arising in the decision process is the presence of uncertain data while decision is made. Recently, this drawback has been treated using stochastic programming techniques in these areas of application [20]. For instance, [30, 48], [7, 53], and [6, 9] propose stochastic programming models to formulate problems related to generators, consumers and retailers, respectively.

In the present chapter we discuss the problem faced by an electricity retailer, which strives to determine a reasonable forward contracting portfolio and competitive selling prices for its clients while preventing itself from uneconomic decisions. See [6], by which the model was inspired and also [25], for more details. We design a two-stage mixed-integer linear program, which—in order to be as competitive as possible—has the minimization of selling prices as objective. To perform well on the (finite) set of scenarios during the planning horizon, the random variable representing the profits of the retailer is enforced to dominate a prespecified profit benchmark profile to second order stochastically. Since obviously, higher profits are preferable to lower profits, we aim at a model of the form (2.14).

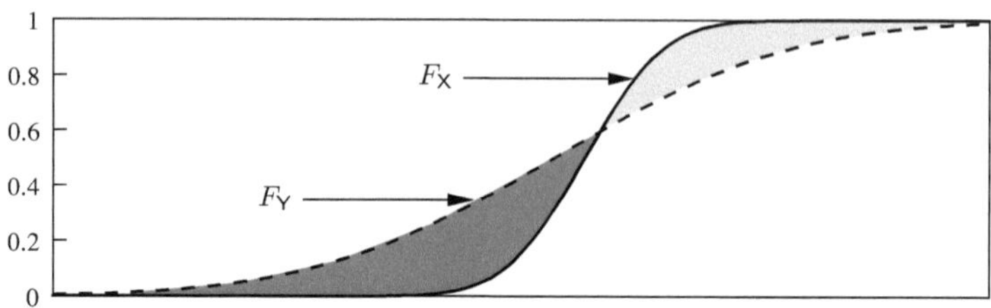

Figure 3.1: Distribution functions of random variables X and Y for which $X \succeq_{(2)} Y$ holds true.

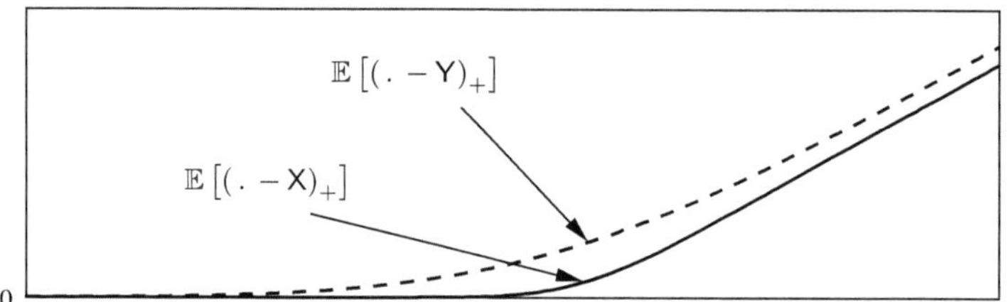

Figure 3.2: Illustration of the defining relation for the distributions in Figure 3.1.

3.2 Retailer Problem with Stochastic Dominance Constraints

3.2.1 Decision Framework

We consider the problem of an electricity retailer that has to decide the forward contracting purchases and the selling prices offered to potential clients at the beginning of the year. In the course of the year, the retailer participates in the pool to supply proportions of it's clients demands or to sell excessive energy previously contracted in the futures market. At the beginning of the year, future pool prices and client demands are unknown for the retailer. Thus the retailer has to decide on the forward contracting portfolio and on selling prices with incomplete information. Therefore, the variables concerning the policies at the beginning of the year (futures market and selling prices) are modeled as *here-and-now* decisions, that have to be taken before knowing the true values of the unknown data (future pool prices and client demands). These unknown parameters are modeled as random vectors. Accordingly, transactions on the pool are modeled as *wait-and-see* variables which depend on the first-stage decisions and the revelation of the random vectors.

decide forwards and prices	$\longrightarrow$	observe pool prices and client demands	$\longrightarrow$	decide transactions with the pool

3.2.2 Forward Contracting

We consider that the retailer participates as a price-maker agent in a futures market where two types of contracts are available: base and peak contracts. Base contracts are available for all 24 hours of the day, while peak contracts are only available for peak hours, i. e., from 11am to 14pm and from 19pm to 21pm.

The price for energy from forwards varies with the quantity purchased. To model this circumstance, we use stepwise constant forward contracting curves similar to that shown in Figure 3.3.

The price $z(P_f)$, the retailer has to pay per hour for the power P_f from contract f is formulated as follows: Let P_f be in block m of the forward curve f, i. e., $P_f \in \left(\sum_{j=1}^{m-1} \bar{P}_{f,j}^C, \sum_{j=1}^{m} \bar{P}_{f,j}^C \right]$ (see Figure 3.3). Then

$$z(P_f) \quad = \quad \sum_{j=1}^{m-1} \bar{P}_{f,j}^C \cdot \bar{\lambda}_{f,j}^C + \left(P_f - \sum_{j=1}^{m-1} \bar{P}_{f,j}^C \right) \cdot \bar{\lambda}_{f,m}^C \qquad (3.1)$$

$$= \quad \bar{\lambda}_{f,m}^C \cdot P_f + \sum_{j=1}^{m-1} \bar{P}_{f,j}^C \cdot \left(\bar{\lambda}_{f,j}^C - \bar{\lambda}_{f,m}^C \right). \qquad (3.2)$$

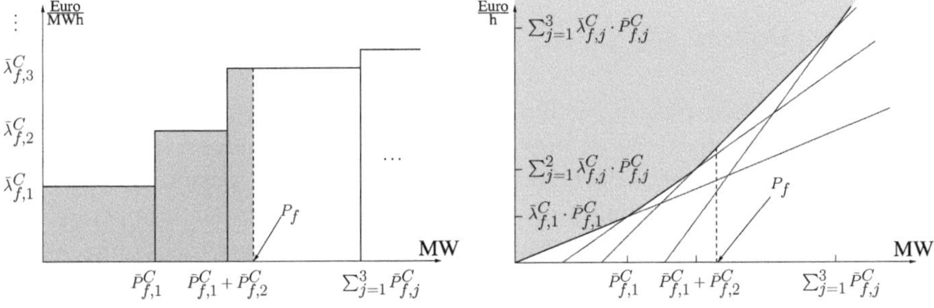

Figure 3.3: First three blocks of a forward contracting curve reflecting the quantity-dependent price for energy.

Figure 3.4: Piecewise linear and convex costs for power from contract f.

The thin lines in Figure 3.4 depict the affine expressions (3.2) for all $m = 1, \ldots, N_B^f$, with N_B^f being the number of blocks of forward contracting curve f.

Altogether $z(\,.\,)$ is a piecewise linear convex cost function. Later on, its minimization will be accomplished by minimizing the bounding variable z_f subject to the N_B^f linear inequalities

$$z_f \geq \bar{\lambda}_{f,m}^C \cdot P_f + \sum_{j=1}^{m-1} \bar{P}_{f,j}^C \cdot \left(\bar{\lambda}_{f,j}^C - \bar{\lambda}_{f,m}^C \right), \ \forall m = 1,\ldots,N_B^f. \qquad (3.3)$$

3.2.3 Pool

The retailer participates in the pool to purchase part of the demand of its clients or to dump excessive energy from contracts. The pool price in each period t is again represented through a set of scenarios $\lambda_{t,\ell}^P$, $\ell = 1,\ldots,N_L$, which can be generated using time series models [27, 28, 29].

Decisions related to the retailer's activity in the pool are the purchase and sale of energy in each period t, denoted as $E_{t,\ell}^P$. The sign of these variables indicates the flow direction.

Since decisions on pool transactions are short-term decisions depending on the forward portfolio and on the actual demands, these variables carry the scenario identifier ℓ.

3.2.4 Client Demand

We consider that the retailer has the potential to supply energy to a set of clients divided into N_E groups (here we implemented three groups: industrial, commercial and residential customers). Each group e is characterized by consumption patterns and its response to the selling price. The total demand of group e in period t is considered to be scenario dependent, whereas the uncertainty that comes along with this fluctuation is of lesser extent than that associated to pool prices.

From the total client demand, the retailer supplies a fraction depending on the selling price offered. The relation between the price offered and the relative end user demand supplied by the retailer is emulated with stepwise constant price quota curves [6]. Figure 3.5 shows the price quota curves we used for the three customer groups under consideration.

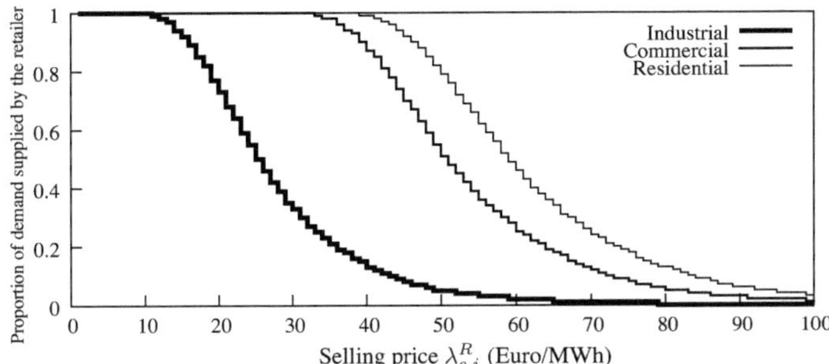

Figure 3.5: Price quota curves.

The supplied demand of end user group e in period t and scenario ℓ, while accounting for (3.5), reads:

$$\sum_{i=1}^{N_I} \bar{E}^R_{t,e,i,\ell} \cdot v_{e,i},\tag{3.4}$$

where the $v_{e,i}$ are binary variables indicating which of the N_I linearity regions $\left[\bar{\lambda}^R_{e,i-1}, \bar{\lambda}^R_{e,i}\right]$ of the price quota curve is active for customer e, hence:

$$\sum_{i=1}^{N_I} v_{e,i} = 1\tag{3.5}$$

$\bar{E}^R_{t,e,i,\ell}$ are the correspondent step heights *times* the customer group's overall demand. From an economic point of view, it is clear, that selling prices should always be a right bound of one of the intervals $\left[\bar{\lambda}^R_{e,i-1}, \bar{\lambda}^R_{e,i}\right]$. Therefore we have that the selling price for customer group e equals:

$$\sum_{i=1}^{N_I} \bar{\lambda}^R_{e,i} \cdot v_{e,i}.\tag{3.6}$$

Independence of the random outcomes of first-stage decisions is a basic assumption in two- and multistage stochastic programming. Note that we just relaxed this restriction. The stochastic client demand depends on the offered prices. Admittedly, in a rather elementary way.

3.2.5 Energy Balance

The end user demand has to be met in each time period of each scenario:

$$E^P_{t,\ell} + \sum_{f \in F_t} P_f \cdot d_t = \sum_{e,i} \bar{E}^R_{t,e,i,\ell} \cdot v_{e,i} \quad \forall t, \forall \ell. \tag{3.7}$$

Pool-sellings ($E^P_{t,\ell} < 0$) and purchases ($E^P_{t,\ell} > 0$) together with forward transactions ($f \in F_t \Leftrightarrow$ contract f is available in period t) enable the retailer to fulfill its commitments in each period of each scenario. Consequently our model features complete recourse. Note, that due to (3.5) the right-hand side of (3.7) contains at most one non-zero summand per customer group, namely the *actual* energy demand procured by the retailer in time interval t.

3.2.6 Profit

The profit of the retailer can be seen as a random variable equal to the revenue obtained from selling electricity to the end users and to the pool minus the cost of purchasing electricity from the pool and through forward trading. The mathematical formulation of the profit in scenario ℓ is:

$$-\sum_t \sum_{f \in F_t} d_t \cdot z(P_f) + \sum_t \left(E^P_{t,\ell} \cdot \lambda^P_{t,\ell} + \sum_{e,i} \bar{E}^R_{t,e,i,\ell} \cdot \bar{\lambda}^R_{e,i} \cdot v_{e,i} \right). \tag{3.8}$$

The costs of purchasing forward contracts is incurred at the beginning of the planning horizon, thus being independent of the scenarios. In contrast, costs and revenues associated with transactions in the pool and sales to end users are dependent on pool price and demand scenarios.

3.2.7 Expectation Model

To have a reference point for the upcoming case study, we firstly set up the purely expectation-based model (1.9). The mixed integer linear programming equivalent of our retailer problem, with π_ℓ denoting the scenario probabilities, then reads:

$$\max \left\{ \quad -\sum_{t}\sum_{f\in F_t} d_t \cdot z_f + \sum_{\ell,t} \pi_\ell \left(E^P_{t,\ell} \cdot \lambda^P_{t,\ell} + \sum_{e,i} \bar{E}^R_{t,e,i,\ell} \cdot \bar{\lambda}^R_{e,i} \cdot v_{e,i} \right) \right.$$

subject to:

Forward contracting constraints: (3.3)

Client demand constraints: (3.5)

Energy balance constraints: (3.7)

$$\left. E^P_{t,\ell}, z_f \in \mathbb{R},\ P_f \in \mathbb{R}_+,\ v_{e,i} \in \{0,1\} \right\} \qquad (3.9)$$

3.2.8 Second Order Stochastic Dominance Model

To be as competitive as possible, we use the sum of the selling prices as the objective of a dominance-constrained retailer problem. Other objective functions are conceivable, e. g., minimal investments in some contracts. According to Corollary 2.17, we state the retailer problem including second order stochastic dominance constraints induced by (here) linear recourse as

$$\min \left\{ \sum_{e,i} \bar{\lambda}^R_{e,i} \cdot v_{e,i} \quad : \quad \begin{array}{l} (3.3),\ (3.5),\ (3.7)\ \text{(The constraints from the}\\ \text{expected value problem (3.9))} \\[1ex] -\sum_t \sum_{f\in F_t} d_t \cdot z_f \\ +\sum_t \left(E^P_{t,\ell} \cdot \lambda^P_{t,\ell} + \sum_{e,i} \bar{E}^R_{t,e,i,\ell} \cdot \bar{\lambda}^R_{e,i} \cdot v_{e,i} \right) \\ +s_{\ell,k} \ \geq\ b_k,\ \forall \ell = 1,\ldots,N_\ell, \\ \forall k = 1,\ldots,N_K \\[1ex] \sum_{\ell=1}^{N_L} \pi_\ell \cdot s_{\ell,k} \ \leq\ \mathbb{E}\left((b_k - B)_+\right), \\ \forall k = 1,\ldots,N_K \\[1ex] E^P_{t,\ell}, z_f \in \mathbb{R},\ P_f, s_{\ell,k} \in \mathbb{R}_+,\ v_{e,i} \in \{0,1\} \end{array} \right\} \qquad (3.10)$$

3.3 Case Study

3.3.1 Data

The performance of the proposed methodology is illustrated by a realistic case study dealing with the electricity market of mainland Spain [86]. This case study is based upon that contained in [6]. For information on the data we used, i. e., how pool prices and end user demands were simulated, we refer the reader to the just mentioned work. The planning horizon of one year was divided into 72 periods as also described in [6] and a set of 200 data scenarios was used, each of which consisting of 72 demand values for each customer group as well as 72 pool prices. Figure 3.6 shows the pool prices for the 72 periods in all of the 200 scenarios. The bold line in Figure 3.6 corresponds to the expected pool prices. The relation between the selling price and the demand provided by the retailer is modeled by the price quota curves from Figure 3.5. It can be observed that industrial consumers are the most sensitive to the price offered by the retailer. Figure 3.7 depicts the demand of the end user groups in each of the 200 scenarios. The bold lines in Figure 3.7 again represent the expected value of the data. The forward contract information is based on data available from the electricity futures market of Spain and Portugal, OMIP [87]. Three monthly and four quarterly contracts each with base and peak hours are considered (14 contracts). Table 3.1 shows the prices of each contract in the first block of the forward contracting curve. Nine additional blocks have been modeled with prices increasing 10% in each additional block. For every contract, the power bounds $\bar{P}^C_{f,j}$ were chosen identical for each block j. The values of $\bar{P}^C_{f,j}$ are listed in columns 3 and 5 of Table 3.1 (cf. Figure 3.3).

Table 3.1: Forward contracting curve parameters.

	Peak		Base	
Contract	Price in first block €/MWh	$\bar{P}^C_{f,j}$ MW	Price in first block €/MWh	$\bar{P}^C_{f,j}$ MW
Monthly 1	42.98	500	31.91	500
Monthly 2	33.57	400	25.29	400
Monthly 3	23.18	200	19.19	200
Quarterly 1	33.24	500	25.47	500
Quarterly 2	30.13	200	25.53	200
Quarterly 3	41.80	100	32.82	100
Quarterly 4	37.32	100	28.87	100

3.3.2 Results

All computations were done with CPLEX 10.0.1, GAMS Link 31 [23, 61] on a Linux-PC with a 2GHz Pentium 4 CPU and 694MB main memory.

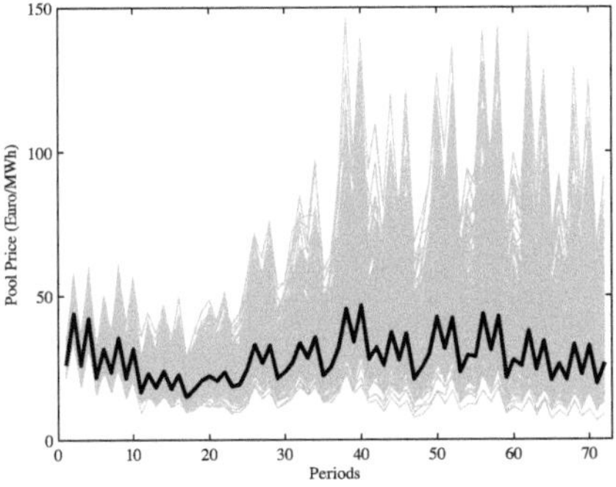

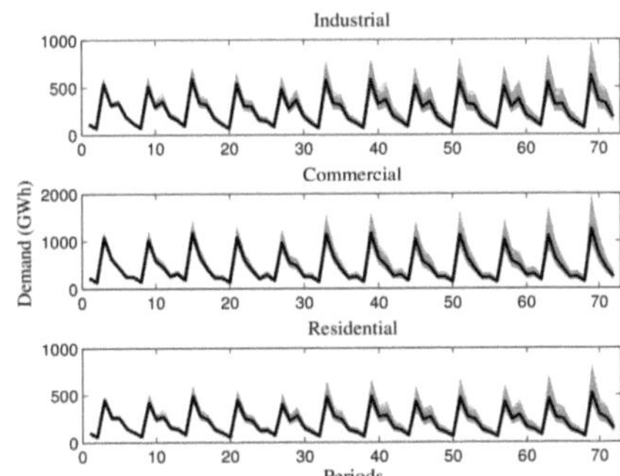

Figure 3.6: Pool prices.   **Figure 3.7:** End user demand.

Expectation Model:

Firstly, we solved the expectation-based problem (3.9). It took six seconds of computing time to reach optimality. The expected yields are $1.038 \cdot 10^9$ €. The probability of losses is 1.3% and the expected shortfall below 0 amounts to $4.001 \cdot 10^6$€. The worst case scenario, having a probability of 0.01%, costs $873.518 \cdot 10^6$€ (losses). The standard deviation of the profits is $507.789 \cdot 10^6$€ (these figures are compiled in the A column of Table 3.2). The A bars in Figure 3.11 visualize the optimal power allocation of the retailer in the futures market (GW/GWh) with respect to maximizing expected yields, which is quite inert in that case. But we will see, that hedging becomes an issue. The resulting selling prices are 40.30, 53.43 and 58.48 €/MWh for industrial, commercial, and residential consumers, respectively.

Benchmark 0:

For a start we chose exactly the profit distribution belonging to the optimal solution x^* of the expected value problem as a benchmark in problem (3.10), i.e., $B(\omega) := \tilde{h}_{x^*}(\omega) \, \forall \omega$ (cf. Corollary 2.17). It is no surprise, that the optimal values of the variables are exactly the same, as after solving the expected value problem. However, in this case the solution time is 957 seconds, which is significantly higher than that needed when solving the expected value problem. Let us remain with this *shape* of the benchmark distribution for a moment, while shifting it slightly to the left and therefore making it easier to dominate it (we now use $B(.) := \tilde{h}_{x^*}(.) - 0.05 \cdot \mathbb{E}(\tilde{h}_{x^*})$). One effect is, that selling prices decrease

Table 3.2: Relevant parameters obtained from solving the models under consideration. RPD stands for *"Resulting profit distribution"* and BD for *"Benchmark distribution"*.

		A	B	C	D
Selling prices	Ind.	40.30	37.27	42.32	50.40
€/MWh	Com.	53.43	53.43	59.49	64.55
	Res.	58.48	54.44	59.49	66.57
Percentage of	Ind.	33	42	27	13
demand covered (%)	Com.	78	78	55	39
	Res.	85	94	82	59
Expected profit	RPD	1.038	0.986	0.917	0.726
(in 10^9 €)	BD	1.0389	0.976	0.915	0.725
Profit standard	RPD	0.508	0.487	0.307	0.136
deviation (in 10^9 €)	BD	0.508	0.508	0.44	0.204
Probability	RPD	1.3%	1.4%	0.03%	0.0%
of losses	BD	1.3%	1.05%	2.2%	0.0%
Expected short-	RPD	−4.001	−4.227	−0.559	0.00
fall below 0 (in 10^6 €)	BD	−4.001	−4.726	−0.658	0.00
Worst case sce-	RPD	0.1%	0.1%	0.1%	0.1%
nario probability	BD	0.1%	0.1%	2.2%	2.2%
Worst case	RPD	−0.874	−0.887	−0.299	0.14
scenario result (in 10^9 €)	BD	−0.874	−0.925	−0.299	0.14
Computing time	RPD	957sec. (6sec.)	3982sec.	644sec.	· 563sec.

(37.27 €/MWh $\downarrow$ 7.5%, 53.43 €/MWh $\downarrow$ 0% and 54.44 €/MWh $\downarrow$ 6.9% for industrial and residential consumers). In addition the purchases in the futures market increase (B bars in Figure 3.11). Consequently the expected profit actually is 5% lower ($0.986 \cdot 10^9$€) compared to that of the preceding two models, where it amounted to $1.038 \cdot 10^9$€. The probability of suffering losses increased by 0.1% to 1.4% and the expected value of shortfalls below 0 is $4.227 \cdot 10^6$€ $\downarrow$ 0.053%. However the profits standard deviation, still being high, is pushed down a little due to investments in the futures market ($487.6 \cdot 10^6$ € $\downarrow$ 3.975%). Figure 3.8 shows the distribution of benchmark 0 (shifted) and the resulting profit distribution when implementing this reference profile into (3.10). As an orientation, we also plotted the distribution of $\tilde{h}_{x^*}$ (the thin solid line).

Figure 3.12(b) symbolizes the mean energy demand of the three customer groups together. The light gray percentage of their mean overall demand is covered by rival retailers. The part supplied by the retailer is subdivided into the part the retailer acquired in the futures market and the part, which is managed using the pool.

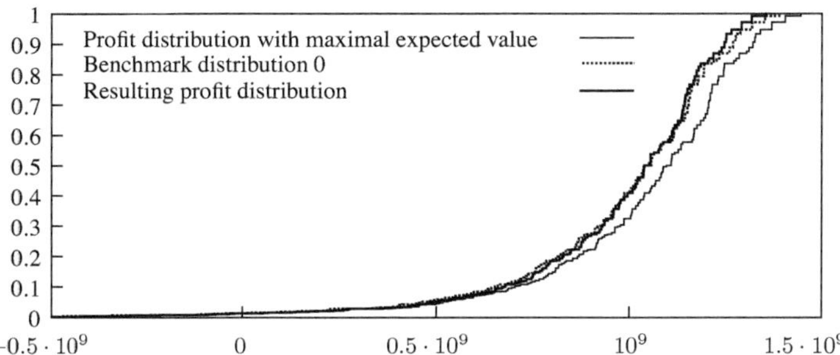

Figure 3.8: Distribution function of $B(\omega) := \tilde{h}_{x^*}(\omega) - 0.05 \cdot \mathbb{E}(f_{x^*}) \; \forall \omega$ (Benchmark) and of the resulting profits when using this B as a benchmark in (3.10). The third graph is the profit distribution resulting from using $B := \tilde{h}_{x^*}$ in (3.10). The same distribution is obtained from solving the expected value problem (3.9). This is always the case, when (3.9) has a unique first-stage solution.

Benchmark 1 and Benchmark 2:

Now we will come to more meaningful choices of benchmark profiles. We define two profit benchmark distributions with 10 scenarios that might reflect economic targets of the retailing company, i.e., the benchmarks feature less standard deviation and less/no losses (see Figures 3.9 and 3.10 as well as Table 3.2). Figures 3.12(c) and 3.12(d) yield the information just described, accordingly for benchmark 1 and benchmark 2. Figures 3.9 and 3.10 show the distribution functions of the employed benchmark profiles and the resulting profit distributions when solving (3.10). The C and D bars in Figure 3.11 show the forward buying when applying benchmark 1 and benchmark 2, respectively. Table 3.2 compiles relevant parameters obtained from solving the models under consideration.

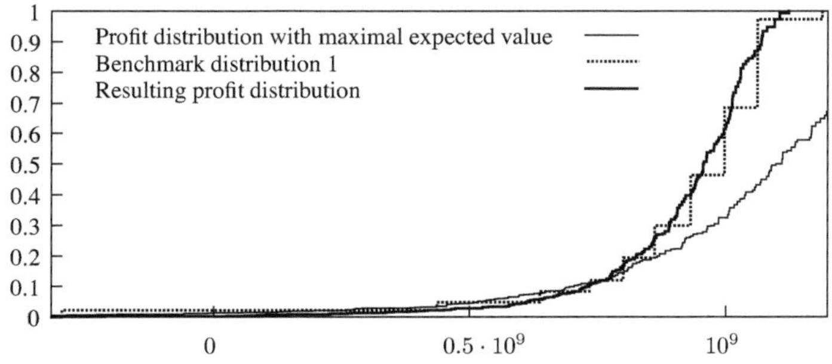

Figure 3.9: Profit distribution function resulting from solving (3.10) while using benchmark 1.

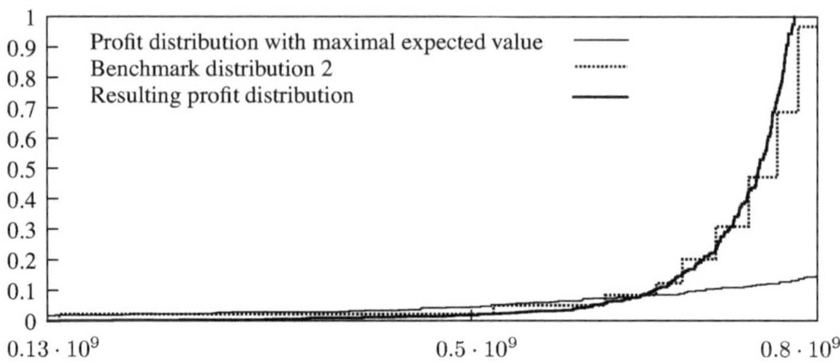

Figure 3.10: Profit distribution function resulting from solving (3.10) while using benchmark 2.

3.4 Conclusions

We presented a new formulation for the problem faced by an electricity retailer that attempts to optimally determine its forward contracting portfolio and the selling price offered to its clients. This problem is formulated as a two-stage stochastic programming problem with second order stochastic dominance constraints. The usage of stochastic dominance constraints in problems formulated from the point of view of electricity market agents is novel and allows the decision-maker to impose its preferences on the resulting profit random variable. The initial stochastic problem has been reformulated as a mixed-integer linear problem through its deterministic equivalent. The proposed formulation has been tested in a realistic case study. The numerical results using different benchmark profiles indicate that the inclusion of stochastic dominance constraints in the formulation of the problem permits an efficient control of the resulting profit variable. The solution times obtained are moderate for different numbers benchmark scenarios. As we will see later, this is not always the case. Especially, when the second stage massively contains integer variables standard solvers fail. To provide a tool to, nevertheless, tackle more complex problems we now come to the development of a decomposition algorithm to solve (2.13) in a different manner, which in many cases is superior to the application of standard software alone.

Space complexity also quickly becomes an issue when addressing (2.13). Sometimes standard software is potentially able to solve instances very efficiently, but the mere size of (2.13) prevents the application due to hardware restrictions and motivates the decomposition approach from the next chapter from a further perspective: While dealing with the investment planning problems for electricity generation from [52] but under increasing convex order constraints, we made the

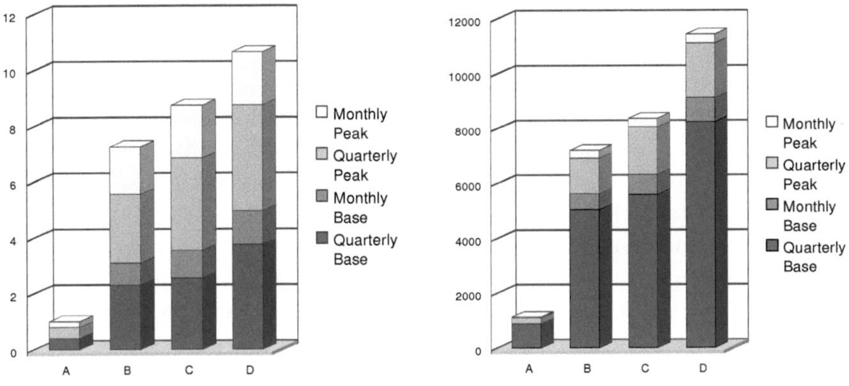

(a) Forward contract purchases in GW.   (b) Forward contract purchases in GWh.

Figure 3.11: These figures visualize the power/energy allocation in the futures market for the models under consideration. The A bars show the buyings for the expected value problem and of the dominance constrained problem when using the profit distribution of the expected value problem as benchmark. B is associated to the buyings when using the profit distribution of the expected value problem, where 5% worse performance with respect to the resulting expected value is tolerated. C and D stand for hedgings when dealing with benchmark 1 and benchmark 2 in (3.10).

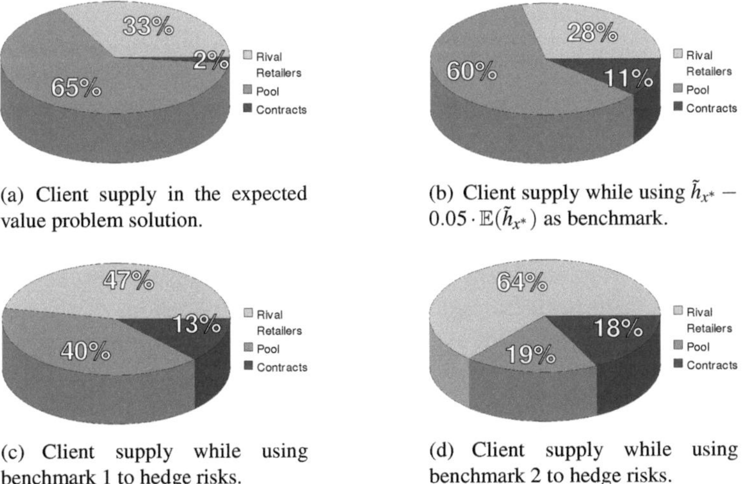

(a) Client supply in the expected value problem solution.

(b) Client supply while using $\tilde{h}_{x^*} - 0.05 \cdot \mathbb{E}(\tilde{h}_{x^*})$ as benchmark.

(c) Client supply while using benchmark 1 to hedge risks.

(d) Client supply while using benchmark 2 to hedge risks.

Figure 3.12: In these figures, the whole pie chart represents the mean overall demand of all customer groups together. The sectors reflect the part of the overall demand supplied using the pool and the signed forward contracts and the part supplied by rival retailers (all mean values).

experience that ILOG CPLEX is able to solve instances within admissible time spans, as long as they fit into the used main memory of 1GB. These energy investment problems that form our second group of test instances are inspired by [73]. We consider two-stage versions of the multi-stage model there and add integrality requirements to the first stage. This leads to a two-stage mixed-integer linear stochastic program where, in the first stage, decisions on capacity expansions for different generation technologies under budget constraints and supply guarantee are made. We assume that these decisions reflect indivisibilities (generation units) and hence are integer-valued. The second stage concerns the minimization of production costs for electricity under the constraints that the electricity demand is met and the available capacity is not exceeded.

The electricity demand is captured by a load duration curve assigning to each duration $\tau \in \mathbb{R}_+$ the minimum load to be covered over time spans adding up to τ. This is where uncertainty enters, since in practice load durations are typically available only stochastically. The model uses step function approximations for load duration curves. So each data scenario is represented by a (finite) step function.

The aim of the optimization is cost minimization where costs are incurred by the expansion decisions of the first stage and the production levels of the second stage. Together with the random load durations this leads to a random optimization problem which is a specification of (1.2).

To derive a benchmark profile a, we first consider $\tilde{f}_{x^*}(\omega)$ where x^* denotes an optimal solution to the expectation model. With heuristically selected benchmark values, the $\tilde{f}_{x^*}(\omega)$ then are clustered around these values, and the probability of each benchmark value arises as the sum of the probabilities of the members in its cluster.

As objective function for the corresponding increasing convex order constrained stochastic program, we consider the capacity expansion of one of the different technologies, possibly one least desired for environmental reasons. The increasing convex order constrained model then minimizes expansions of this capacity over all expansion policies, not exceeding the benchmark profile in terms of this stochastic order.

CPLEX (with default settings) was able to solve instances 1.1 and 2.1 from Table 3.3 in the root node after applying some heuristics and mixed integer rounding cuts. For instances 1.2 and 2.2 the provided main memory was insufficient and CPLEX was aborted by the operating system without having found a feasible point. Later, in Table 4.1 it will become clear, that the decomposition, requiring much less main memory than plain CPLEX does for tackling deterministic equivalents, is able to solve all instances from Table 3.3.

Table 3.3: Results from the application of CPLEX to deterministic equivalents on standard PC with 1GB RAM.

	Benchmarks		CPLEX 9.1.3		
	Probability $\mathbb{P}(a = a_k)$	Benchmark values $a_k, k = 1, \ldots, K$	Time (sec.) / Status	Lower Bound	Upper Bound
Instance 1.1 with 5000 data scenarios	0.203 0.198 0.2 0.2 0.199	75.597 78.853 81.76 85.31 101.8	445 Status: Integer optimal	108	108
Instance 1.2 with 5500 data scenarios	0.203 0.198 0.2 0.2 0.199	75.597 78.853 81.76 85.31 101.8	278 Status: Out of memory	106.974	–
Instance 2.1 with 5000 data scenarios	0.203 0.198 0.2 0.2 0.199	78.097 81.353 84.26 87.810 104.3	856 Status: Integer optimal	0	0
Instance 2.2 with 5500 data scenarios	0.203 0.198 0.2 0.2 0.199	78.097 81.353 84.26 87.810 104.3	2697 Status: Out of memory	0	–

Chapter 4

Decomposition Method

Recall that

$$\min \left\{ g^\top x \ : \ \tilde{f}_x \leq_{icx} a \ , \ x \in X \right\}$$

$(|\tilde{f}_x(\Omega)| = L < \infty, \ |a(\Omega)| = K < \infty)$ was shown to be equivalent to

$$\min \left\{ g^\top x : \begin{array}{lll} c^\top x + q^\top y_\ell - v_{\ell k} & \leq a_k & \forall \ell \quad \forall k \\ Tx + Wy_\ell & = z_\ell & \forall \ell \\ \sum_{\ell=1}^{L} \pi_\ell v_{\ell k} & \leq \mathbb{E}[(a - a_k)_+] & \forall k \\ x \in X, \ y_\ell \in Y, & v_{\ell k} \geq 0 & \forall \ell \quad \forall k \end{array} \right\}, \qquad (4.1)$$

where $\ell = 1, \ldots, L$ and $k = 1, \ldots, K$, in the sense that the optimal values coincide
and that its arguments are feasible for both problems.

As a start we formulate a generic branch-and-bound algorithm for (4.1), by which
the set X is partitioned with increasing granularity. To maintain the mixed-integer
linear problem formulation, linear inequalities are used for this partitioning. On
the current elements of the partition, upper and lower bounds for the optimal objective function value are computed. This is embedded into a coordination procedure
(cf. Figure 4.1) to guide the partitioning and to prune elements due to infeasibility, optimality or inferiority. Altogether, tighter and tighter bounds for the global
optimal value are generated. A decomposition effect will come up in the upper-
and lower bounding subroutines of the algorithm, that will be discussed in what
follows.

By **P** we denote a list of problems, and $\varphi_{LB}(P)$ is a lower bound for the optimal value of $P \in \mathbf{P}$. Moreover, $\bar{\varphi}$ denotes the currently best upper bound to the
optimal value of (4.1), and $X(P)$ is the element in the partition of X belonging to P.

Algorithm 1: Computing a Solution of (4.1)

  1: **Input: P** $:= \{(4.1)\}$
  2: **Output:** Solution of (4.1), or detection of infeasibility or unboundedness
  3: **initialize** $\bar{\varphi} \in \mathbb{R} \cup \{\pm\infty\}$ by $+\infty$

④: **while** $\mathbf{P} \neq \emptyset$ **do**
  5:      select and delete a problem P from $\mathbf{P}$
  6:      compute a lower bound $\varphi_{LB}(P)$
  7:      **if** $\varphi_{LB}(P) \in [\bar{\varphi}, +\infty]$ (inferiority of P or infeasibility of a subproblem) **then**
  8:          **go to** (line) ④
  9:      **else**
10:         apply a feasibility heuristic to find a feasible first stage $\bar{x}$ of P yielding an upper bound $g^\top \bar{x}$ for P.
11:     **end if**
12:     **if** $g^\top \bar{x} < \bar{\varphi}$ **then**
13:        $\bar{\varphi} := g^\top \bar{x}$
14:     **end if**
15:     **if** $\varphi_{LB}(P) = g^\top \bar{x}$ (optimality for P) **then**
16:        **go to** ④
17:     **end if**
18:     create two new subproblems by partitioning the set $X(P)$ by means of linear inequalities (cf. Remark 4.1). Add these subproblems to $\mathbf{P}$ and **go to** ④
19: **end while**
20: **return** $\bar{x}$ that yielded $\bar{\varphi} = g^\top \bar{x}$ (the y- and v-variables are also known)

Of course, Algorithm 1 is of little value as long as the bounding procedures invoked in lines 6 and 10 are not specified.

Let us start with lower bounding. The basic idea is to pass to a model in L-shaped form by means of relaxation. In view of the discussion in Section 2.2 the obvious candidates for this relaxation are $\sum_{\ell=1}^{L} \pi_\ell v_{\ell k} \leq \mathbb{E}[(a - a_k)_+]$, $k = 1,\dots,K$ (cf. (2.12)). The number K of realizations of a originates from a subjective perception of risk, and is often quite small, say at most within some tens, compared with the generally far bigger number L of data scenarios. Hence, Lagrangean relaxation of these constraints will lead to a Lagrangean dual of tractable dimension.

For models in L-shaped form, two principal decomposition approaches can be taken, a Benders-type decomposition or a dual decomposition based on (Lagrangean) relaxation of nonanticipativity [20, 24, 65, 94, 99, 100, 112]. With integer variables in the second stage, however, Benders decomposition leads to nonconvex master problems. Therefore, we pursue dual decomposition. We relax nonanticipativity of x in problem (4.1) by introducing copies x_ℓ, $\ell = 1,\dots,L$ of x and of all second-stage constraints containing nonanticipative variables. This transforms the constraint matrix of (4.1) (cf. Figure 2.5) into a new matrix with exploitable structure which is shown in Figure 4.2. One possibility now could

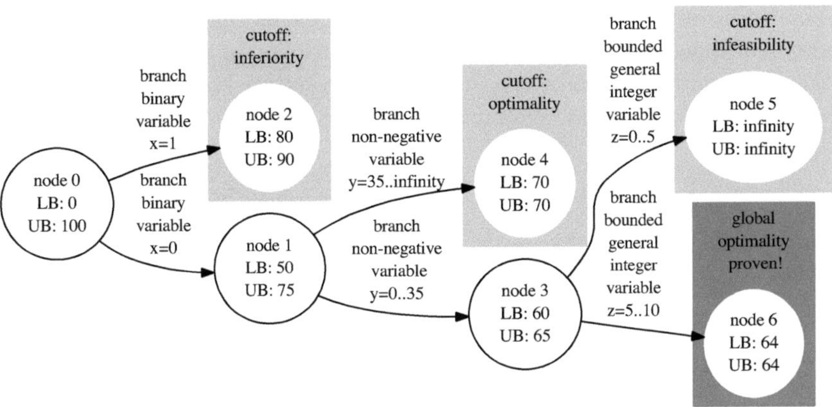

Figure 4.1: Visualization (part of the output of our implementation) of a conceivable cycle of Algorithm 1. LB and UB stand for lower and upper bounds in the current node respectively. The evolution of $\bar{\varphi}$ is $100, 75, 75, 65, 65, 65, 64$. That of the global lower bound is $0, 50, 50, 60, 60, 60, 64$.

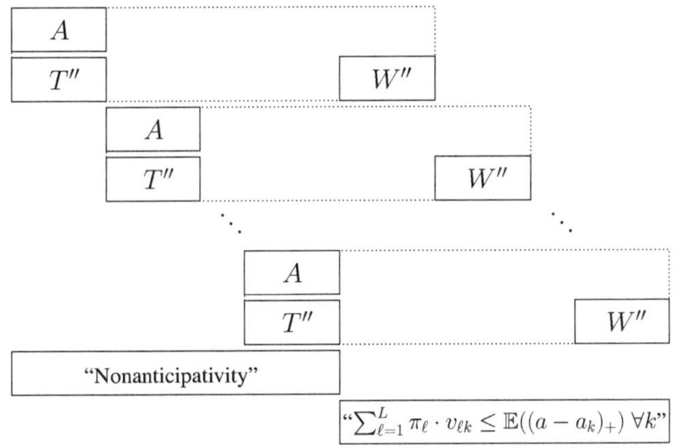

Figure 4.2: Structure of (4.1) after cloning the first-stage variables and constraints and after introducing explicit nonanticipativity constraints.

be to regain nonanticipativity by Lagrangean relaxation of some representation of the identities $x_1 = x_2 = \ldots = x_L$[17]. This, however, would lead to a Lagrangean

---

[17] "Nonanticipativity" $\widehat{=}$ $\boxed{H_1}\ \ldots\ \boxed{H_L}$ $\in \{-1, 0, 1\}^{((L-1)\cdot\dim X)\times(L\cdot\dim X)}$, with e. g., $(H_1)_{\bullet j} := (\bar{0}_{(\dim X\cdot(j-1))}, \bar{1}_{(\dim X)}, \bar{0}_{(\dim X\cdot(L-j-1))})^\top$ ($\bar{1}$ is defined analogously to $\bar{0}$, cf. Figure 2.4) and $(H_i)_{\bullet j} := (\bar{0}_{(\dim X\cdot(j-1)+i-2)}, -1, \bar{0}_{(\dim X\cdot(L-j)-i+1)})^\top$, $i = 2, \ldots, L$

dual in dimension $(L-1)\cdot \dim X$, which quickly can become several tens or even hundreds of thousands. Therefore, we leave it at working with the copies x_ℓ in our lower bounding scheme at first, striking a compromise between computational effort and quality of bounds. With these presuppositions, and putting

$$x = \sum_{\ell=1}^{L} \pi_\ell x_\ell \,, \quad \lambda := (\lambda_1,\ldots,\lambda_K)^T \in \mathbb{R}_+^K \,, \quad \Delta := (v_{\ell k})_{\substack{\ell=1,\ldots,L \\ k=1,\ldots,K}}$$

we arrive at the following Lagrangean function (see, e. g., [81, 101])

$$\mathscr{L} : X \times \mathbb{R}_+^{L \times K} \times \mathbb{R}_+^K \longrightarrow \mathbb{R}$$

$$(x_1,\ldots,x_L,\Delta,\lambda) \longmapsto \sum_{\ell=1}^{L} \pi_\ell \cdot \mathscr{L}_\ell(x_\ell,\Delta_{\ell\bullet},\lambda),$$

where $\Delta_{\ell\bullet}$ denotes the ℓ-th row of Δ and

$$\mathscr{L}_\ell(x_\ell,\Delta_{\ell\bullet},\lambda) := g^T x_\ell + \sum_{k=1}^{K} \lambda_k \cdot (v_{\ell k} - \mathbb{E}\left[(a-a_k)_+\right]) \,,$$

for $\ell = 1,\ldots,L$.

It can easily be shown that the solution of (4.1) with $g^\top x$ replaced with the Lagrangean $\mathscr{L}(x_1,\ldots,x_L,\Delta,\lambda)$ yields a lower bound for the original problem for each fixed $\lambda \geq 0$.[18] Since we are keeping track of finding a lower bound as large as possible, we are interested in the solution of the Lagrangean dual. This amounts to $\max\{\mathscr{D}(\lambda) : \lambda \geq 0\}$, where

$$\mathscr{D}(\lambda) := \min \left\{ \mathscr{L}(x_1,\ldots,x_L,\Delta,\lambda) : \begin{array}{l} c^T x_\ell + q^T y_\ell - v_{\ell k} \leq a_k \\ Tx_\ell + Wy_\ell = z_\ell \\ v_{\ell k} \geq 0 \,, \; x_\ell \in X \,, \; y_\ell \in Y \\ k = 1,\ldots,K \,, \; \ell = 1,\ldots,L \end{array} \right\}$$

This is where decomposition becomes effective. The optimization problem behind $\mathscr{D}(\lambda)$ is separable in ℓ, and we obtain

---

[18] Consider $\hat{x} := \arg\min\{c^\top x : Ax \leq b \,, x \geq 0\}$ and $\mathscr{D}(\lambda) := \min_{x\geq 0}\{c^\top x + \lambda^\top (Ax - b)\}$. For fixed x, $c^\top x + \lambda^\top (Ax - b)$ is affine in λ. Hence, $\mathscr{D}$, as the pointwise minimum of these functions is piecewise linear and concave. Its negative is convex and subdifferentiable. For all $\lambda \geq 0$ it holds $c^\top \hat{x} \geq c^\top \hat{x} + \underbrace{\lambda^\top (A\hat{x} - b)}_{\leq 0} \geq \mathscr{D}(\lambda)$.

$$\mathscr{D}(\lambda) \;=\; \sum_{\ell=1}^{L} \pi_\ell \cdot \min \left\{ \mathscr{L}_\ell(x_\ell, \Delta_{\ell\bullet}, \lambda): \;\; \begin{array}{l} c^\top x_\ell + q^\top y_\ell - v_{\ell k} \le a_k \\ T x_\ell + W y_\ell = z_\ell \\ x_\ell \in X,\; y_\ell \in Y,\; v_{\ell k} \ge 0 \\ k = 1,\ldots,K,\; \ell = 1,\ldots,L \end{array} \right\} \tag{4.2}$$

The function $\mathscr{D}(.)$ is piecewise linear and concave. So bundle methods for nonsmooth convex minimization can be employed for solving the Lagrangean dual, whose optimal value provides a lower bound for the optimal value of (4.1). A simple but very slowly converging alternative would be the iterative method ([5, 91, 92])

$$\lambda_{n+1} := \lambda_n - s_n \cdot \lambda_n^*$$

with

$$\lambda_n^* \;\in\; \partial(-\mathscr{D}(\lambda_n)) \quad (\partial \text{ denotes the subdifferential})$$

$$=\; \operatorname*{conv}_{(x^*,\Delta^*)\in \arg\min\{\mathscr{D}(\lambda_n)\}} \left\{ \left[-\sum_{\ell=1}^{L} \pi_\ell \cdot \left(v_{\ell k}^* - \mathbb{E}((a - a_k)_+) \right) \right]_{k=1}^{K} \right\}$$

and the step size s_n satisfying

$$s_n \to 0, \; \sum_{n=1}^{\infty} s_n = \infty.$$

In our numerical experiments we have used Christoph Helmberg's implementation of the spectral bundle method from [59]. We also considered (and implemented) another alternative Lagrangean function, where at least some of the explicit nonanticipativity restrictions, namely the "most violated" ones, are treated with Lagrangean relaxation. Let $n_1, \ldots, n_M$, $M \in \mathbb{N}$, $M \le \dim X$ denote the indices of the components of the scenario solutions x_ℓ which yield the M largest values of

$$\max_{\ell=1,\ldots,L} \{x_{\ell j}\} - \min_{\ell=1,\ldots,L} \{x_{\ell j}\}. \tag{4.3}$$

We call this value "the" dispersion norm of the component x_j. Then, with $\mu \in \mathbb{R}^{M \times (L-1)}$ we consider the Lagrangean

$$\begin{aligned} \mathscr{L}(x_1,\ldots,x_\ell,\Delta,\lambda,\mu) \;:=&\; \mathscr{L}(x_1,\ldots,x_L,\Delta,\lambda) \\ &+ \sum_{\ell=2}^{L} \sum_{m=1}^{M} \mu_{(\ell-1)m} \cdot \pi_\ell \cdot (x_{1 n_m} - x_{\ell n_m}). \end{aligned}$$

To preserve the decomposition effect as in (4.2) it is necessary to rearrange the double sum as follows:

$$\sum_{\ell=2}^{L} \sum_{m=1}^{M} \mu_{(\ell-1)m} \cdot \pi_{\ell} \cdot (x_{1n_m} - x_{\ell n_m})$$

$$= \pi_1 \sum_{m=1}^{M} \left(\sum_{\ell'=2}^{L} \mu_{(\ell'-1)m} \cdot \frac{\pi_{\ell'}}{\pi_1} \right) \cdot x_{1n_m} - \sum_{\ell=2}^{L} \pi_{\ell} \sum_{m=1}^{M} \mu_{(\ell-1)m} \cdot x_{\ell n_m}.$$

This yields

$$\mathscr{L}(x_1,\ldots,x_\ell,\Delta,\lambda,\mu) := \sum_{\ell=1}^{L} \pi_\ell \cdot \tilde{\mathscr{L}}_\ell(x_\ell,\Delta_{\ell\bullet},\lambda,\mu), \tag{4.4}$$

with the summands

$$\tilde{\mathscr{L}}_\ell(x_\ell,\Delta_{\ell\bullet},\lambda,\mu) := \mathscr{L}_\ell(x_\ell,\Delta_{\ell\bullet},\lambda) + \sum_{m=1}^{M} v_{\ell m} \cdot x_{\ell n_m},$$

where

$$v_{\ell m} := \begin{cases} \sum_{\ell'=2}^{L} \mu_{(\ell'-1)m} \cdot \frac{\pi_{\ell'}}{\pi_1} & , \text{ for } \ell = 1 \\ -\mu_{(\ell-1)m} & , \text{ for } \ell \geq 2, \end{cases}$$

having the desired decomposition feature. Using the adapted Lagrangean (4.4), unmet constraints $\sum_{\ell=1}^{L} \pi_\ell v_{\ell k} \leq \mathbb{E}[(a - a_k)_+]$ are penalized as before and in addition "rough" violations of the $M \cdot (L-1)$ explicit nonanticipativity constraints $x_{1n_m} - x_{\ell n_m} = 0$, $m = 1,\ldots,M$, $\ell = 2,\ldots,L$ are treated with Lagrangean relaxation. Note that μ underlies no sign restrictions as here, opposed to the first group of relaxed constraints, deviations from equality to both sides have to be penalized.

Algorithm 2: Heuristic (Computing an upper bound for (4.1))

1: **Input:** The vectors x_ℓ, $\ell = 1,\ldots,L$ of first-stage decisions from the lower bounding procedure just described
2: **Output:** An upper bound for the current node, or just the information, that the current node has to be branched further
3: Understand x_ℓ, $\ell = 1,\ldots,L$ as suggestions for x and use these to derive a single promising candidate $\bar{x}$ for (4.1). For instance, this can be done by averaging and rounding where required (cf. [54], where lots of details regarding the implementation of the algorithms are explained).
4: **for** $\ell = 1$ to L **do**

5: **if**

$$\min\left\{ \sum_{k=1}^{K} v_{\ell k} : \begin{array}{ll} c^\top \bar{x} + q^\top y_\ell - v_{\ell k} & \leq a_k \\[1em] T\bar{x} + W y_\ell & = z_\ell \\[1em] y_\ell \in Y, \ v_{\ell k} \geq 0, & k = 1,\dots,K \end{array} \right\}$$

is infeasible **then**

6:      Heuristic stops with the formal upper bound $+\infty$

7:    **end if**

8: **end for**

9: **if** the $v_{\ell k}$ from the preceding step fulfill

$\sum_{\ell=1}^{L} \pi_\ell \cdot v_{\ell k} \leq \mathbb{E}((a - a_k)_+) \ \forall k = 1,\dots,K$ **then**

10:     Feasible point found! Heuristic stops with the upper bound $g^\top \bar{x}$.

11: **else**

12:     Heuristic stops with the formal upper bound $+\infty$.

13: **end if**

Remark 4.1. To preserve a binary structure in the outer branch and bound tree on the one hand and mixed integer linear programming formulations on the other, the branching step (line 18 of Algorithm 1) is carried out in the following way: We successively subdivide subsets of X (beginning with X itself) by means of *two linear inequalities*. For this purpose, we select a component x_j of x having maximal range according to (4.3) among the scenario solutions $(x_\ell)_{\ell=1}^{L}$. More sophisticated dispersion measures are thinkable, e. g., those, taking relative dispersions into consideration. Let x_j be the identified component. Then we obtain the two new problems from the current P, that will be added to the list $\mathbf{P}$ by adding the constraints $x_j \leq \lfloor \bar{x}_j \rfloor$ and $x_j \geq \lfloor \bar{x}_j \rfloor + 1$, respectively if x_j has to be integral or $x_j \leq \bar{x}_j - \varepsilon$ and $x_j \geq \bar{x}_j + \varepsilon$, respectively, where $\varepsilon > 0$ is a tolerance parameter to have disjoint subdomains.

Remark 4.2. Suppose that X is bounded and that some stopping criterion is used to avoid endless branching on continuous components of x, e. g., the value of (4.3) is set to 0 for continuous variables if it falls below some tolerance threshold. In other words this corresponds to cease the branching process at the latest when the subdomain's ℓ_∞-diameter falls below some ε'. Hence the established branch and bound algorithm terminates in finitely many steps and tends to the correct solution, when the first stage is integral. However it is clear that problem (2.13) is NP-hard [22].

Remark 4.3. In line 5 of Algorithm 2 feasibility of $\bar{x}$ for data scenario ℓ is checked. If $\bar{x}$ is infeasible for one of these subproblems, $\bar{x}$ is dropped. The purpose of the objective function $\sum_{k=1}^{K} v_{\ell k}$ is to "push down" the $v_{\ell k}$ in order to fulfill the inequality in line 9. The question now is, how this can be justified, since we minimize the sum of the $v_{\ell k}$ over k while for feasibility of the inequalities in line 9, a weighted sum over ℓ (the expectation of the $v_{\ell k}$) is of crucial importance. In other words: Why are the $v_{\ell k}$ chosen optimally with respect to being small in the increasing convex order? However, again the special structure of (4.1) obliges us: The problems in line 5 decouple with respect to k, i. e., it makes no difference to the solution vector, if we compute

$$
\min\left\{\sum_{k=1}^{K} v_{\ell k} : \ldots\right\} \quad \text{or} \quad \sum_{k=1}^{K} \min\left\{v_{\ell k} : \ldots\right\}. \tag{4.5}
$$

Hence every single $v_{\ell k}$ is chosen as small as possible. Consequently, after successful termination, our algorithm immediately yields a member of $(\tilde{f}_x)_{x \in X}$ (keyword *cost-optimal second stage*, cf. Section 2.2, p. 28).

Sometimes the lower bounding procedure can be shortened. For instance in line 7 of Algorithm 1 the resulting lower bound is used to test the current region of X for infeasibility or inferiority. This can already be done after each descent step of the bundle method. As soon as $\mathcal{D}(\lambda)$ is larger than $\bar{\varphi}$ the current node can be cut off. Similarly, if a "trivial" upper bound for the original problem is known $\mathcal{D}(\lambda)$ can be compared to this bound in each iteration. If $\mathcal{D}(\lambda)$ is greater than this bound, the current node can also be fathomed, before the dual method terminates.

If $g^\top x$ is integral by nature, the result $\varphi_{LB}(P)$ of the lower bounding procedure can be rounded to the next integer.

Finally we mention a specific feature of the decomposed model.

Remark 4.4. $\mathcal{D}(\lambda)$ is a lower bound for the full problem (2.13). Same for $\lambda = \bar{0}_K$, but $\mathcal{D}(\bar{0}_K)$, which is an expected value, can often be tightened significantly in a manner not working for the decomposition of the classical expectation-based model. It holds that

$$
\max_{\ell=1,\ldots,L}\left\{\min\left\{g^T x_\ell : \begin{array}{l} c^T x_\ell + q^T y_\ell - v_{\ell k} \leq d_k \\[4pt] T x_\ell + W y_\ell = z_\ell \\ v_{\ell k} \geq 0, \ x_\ell \in X, \ y_\ell \in Y \\ k = 1,\ldots,K \end{array}\right\}\right\} \tag{4.6}
$$

is a lower bound for the full problem (2.13). In other words, instead of the expected value behind $\mathcal{D}(\bar{0}_K)$, the largest contributor, irrespective the scenario probability, can be taken as a valid lower bound. To show this, let ℓ^* be the index of the subproblem with the maximal objective value. Assume there was a x^* feasible for (2.13) with: $g^T x^* < g^T x_{\ell^*}$. Because x^* is also feasible for subproblem ℓ^*, this is a contradiction to $g^T x^* < g^T x_{\ell^*}$. That is, the validity of this bound is a consequence of the fact that all subproblems in (4.6) are relaxations of the full problem. The objective is the same and the set of constraints is diminished. For the dual decomposition of $\min_{x \in X}\{\mathbb{E}(f_x)\}$ the situation is different due to second-stage variables in the objective, which might yield arbitrarily high scenario costs which are pushed down by low probabilities in the overall objective. Actually, this, say "quick-and-dirty" method, does not require dual iterations and is often preferable to the time consuming approximation of $\max_{\lambda \geq 0}\{\mathscr{D}(\lambda)\}$.

To conclude this chapter we return to the motivating example from the end of Chapter 3. Table 4.1 displays the results of the application of the decomposition method to the four instances from Table 3.3 (p. 47) while using the bounding procedure from Remark 4.4. The three right columns review the results form Table 3.3.

Table 4.1: Comparison of results from solving decomposed models with our implementation ddsip.vSD and from solving deterministic equivalents with plain CPLEX on standard PC with 1GB RAM.

	ddsip.vSD			CPLEX 9.1.3		
	Time (sec.) (evolution)	Lower bound	Upper bound	Time (sec.) / Status	Lower Bound	Upper Bound
Instance 1.1	886	106	341	445	108	108
with	5045	106	212	Status:		
5000	8291	106	141	Integer		
data	12089	106	108	optimal		
scenarios	146759	108	108			
Instance 1.2	1306	106	341	278	106.974	–
with	5812	106	212	Status:		
5500	9264	106	141	Out of		
data	11244	106	108	memory		
scenarios	227127	108	108			
Instance 2.1	26	0	199	856	0	0
with	4680	0	100	Status:		
5000	7120	0	50	Integer		
data	8430	0	25	optimal		
scenarios	10158	0	0			
	To be continued on the next page					

	ddsip.vSD			CPLEX 9.1.3		
	Time (sec.) (evolution)	Lower bound	Upper bound	Time (sec.) / Status	Lower Bound	Upper Bound
Instance 2.2	66	0	199	2697	0	–
with	5279	0	100	Status:		
5500	8016	0	50	Out of		
data	9448	0	25	memory		
scenarios	11309	0	0			

Remark 4.5. (Linear Recourse.) The retailer problem from Chapter 3 and the energy investment problem from the very end of Chapter 3 feature linear recourse. When we reviewed results concerning the problem class

$$\min_{x \in X} \left\{ g(x) : \tilde{f}_x \leq_{st} a \right\},$$

we mentioned the work [40], where a cutting-plane algorithm that exploits the linear recourse property is proposed. However the ideas presented in [40] cannot analogously be translated into the $\leq_{icx}$-setting. The reason for that is, that Proposition 2.8 from [40] is not transferable.
Precisely

$$\min_{x \in X} \left\{ g^\top x : \tilde{f}_x \leq_{icx} a \right\}$$

is *not* equivalent to[19]

$$\min \left\{ g^\top x : \begin{array}{lll} (z_\ell - Tx)^\top \delta_i + (c^\top x - a_k)\delta_{io} - v_{\ell k} & \leq 0 & \forall \ell \; \forall k \; \forall i \\ \sum_{\ell=1}^{L} \pi_\ell v_{\ell k} & \leq \mathbb{E}[(a - a_k)_+] & \forall k \\ & x \in X, v_{\ell k} \geq 0 & \forall \ell \; \forall k \end{array} \right\},$$

where (δ_i, δ_{io}) denote the vertices of

$$\left\{ (u, u_0) : 0 \leq u \leq 1 \text{ (componentwise)}, \; 0 \leq u_0 \leq 1, \; W^\top u - q u_0 \leq 0 \right\}.$$

Example 4.6. Choosing $c = 1$, $q = -2$, $A = 1$, $T = (0,1)^\top$, $W = (-1,-1)^\top$ and $z(\omega) = (\xi(\omega), 0)^\top$ in

$$\min_{x,y \geq 0} \left\{ c^\top x - q^\top y : Ax \leq b, \; Tx + Wy \geq z \right\}$$

---

[19]The problems arise during the adapted proof of $S_2 \subset S_1$ (in the terminology of [40]) because the feasibility problems only yield the information whether $c^\top x + \Phi(z - Tx)$ exceeds a certain value or not, but not to which extent.

yields

$$\min_{x,y\geq 0} \left\{ x - 2y : x \leq 100 , \; -y \geq \xi , \; x - y \geq 0 \right\}.$$

With uniformly distributed data scenarios $z_1 = (-50,0)^\top$, $z_2 = (-100,0)^\top$, uniformly distributed benchmark scenarios $a_1 = -75$, $a_2 = -25$ and $g^\top x = x$, $\min_{x \in X} \left\{ g^\top x : \tilde{f}_x \leq_{icx} a \right\}$ is equal to 50 while the objective value of the above problem collecting all the cuts is 25.

However other (standard) cutting-plane methods are thinkable.

Chapter 5

Test Instances

In the following, we report computational results for Algorithm 1 applied to test instances from power planning and Sudoku puzzling. The first group of instances refers to the optimal management of a dispersed generation system including investment decisions. Similar models have also been dealt with in [50, 55, 56]. The instances of the second group are inspired by [64], see also [50].

5.1 Optimal Planning of a Local Heat Network under Delivery Commitment and Preference of Certain Technologies

5.1.1 Expectation-Based Approach

The energy supply (heat and electricity) of a residential area shall be planned in a way that the sum of investment costs and expected production costs is minimized. Some houses are potential sites for combined heat and power producing (CHP) units. If a CHP unit is installed, it is necessary to distribute excessive heat through a local pipework to meet the demand of all residents. Excessive electric energy can be fed into the main grid against payment. In the first stage the topology of the distribution network and the investment on CHP units has to be decided. The actual energy demand becomes evident in the second stage and has to be met by the provided infrastructure in each scenario (see Figure 5.1).

5.1.2 Data

Load profiles, reflecting the variation in the electrical and thermal load versus time for each house were sampled from realistic demand curves for the whole settlement for a whole year. These were taken from [109] for electric demand. Thermal demand was simulated according to the results achieved in [58] using data from [38]. The sampling was done such that the overall demand is conserved. A scenario is formed by the thermal and electric demand values for each house in one randomly selected hour out of the 8760 hours of the year.

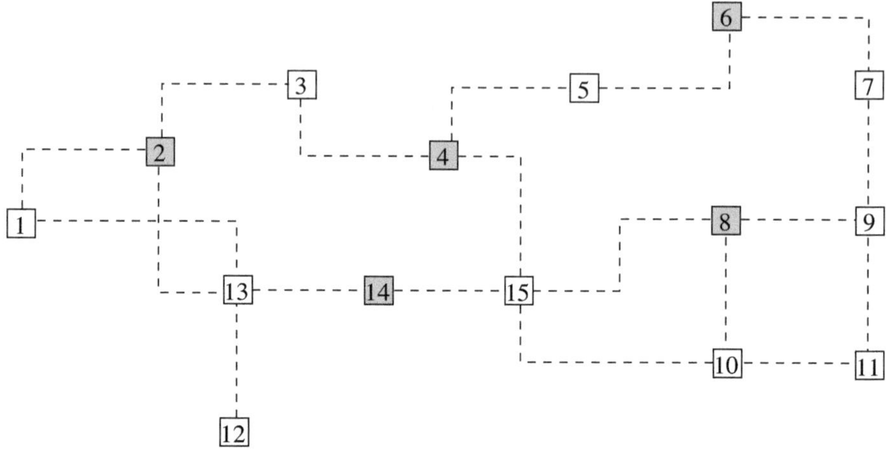

Figure 5.1: Residential area (schematically)

5.1.3 Constants and Variables

Table 5.1: Constants in the Model

Notation	Meaning	Unit
H	Number of houses	-
N	Number of different CHP technologies	-
$S \subset \{1,\dots,H\}$	Set of possible locations for CHP units	-
$\bar{S}$	Complement of S	-
$E \subset \{1,\dots,H\}^2$	Possible locations for transmission lines	-
$\bar{E}$	Complement of E	-
$norm$	1- or 2-norm	-
$rate[1..N]$	Rate for CHP units	%
$lifetime[1..N]$	Lifetime CHP units	a
$Inv\_costs[1..N]$	Annualized investment costs CHP units	€
$price[1..N]$	Investment costs units	€
$rate\_netw$	Rate network	%
$lifetime\_netw$	Lifetime network	a
Inv_costs_netw	Annualized investment costs network	€

To be continued on the next page

Notation	**Meaning**	**Unit**
$price\_netw$	Investment costs network per meter	€
$xy\_pos[1..H, 1..2]$	Location of houses	-
$dist[1..H, 1..H]$	Distances according to parameter $norm$	m
$\eta_{el}[1..N]$	Electric efficiency CHP units	-
$a[1..N]$	Ratio of thermal and electric output	-
C	Fuel cost per unit	€
C_p	Price purchased electricity per unit	€
C_s	Income sold electricity per unit	€
$C_{O\&M}$	Operating and maintenance costs per produced unit	€
$D_{th}[1..H]$	Thermal demand	kW
$D_{el}[1..H]$	Electric demand	kW
$P_{th,netw,abs,ub}$	Upper bound for transmission	kW
$P_{minmax}[1..N, 1..2]$	Minimal/maximal electric output	kW
$P_{th,minmax}[1..N, 1..2]$	Minimal/maximal thermal output	kW

Table 5.2: Variables in the Model

Notation	**Meaning**	**Range**
First-stage variables:		
$Inv\_node[1..H, 1..N]$	Investment decision units	$\{0,1\}$
$Inv\_edge[1..H, 1..H]$	Investment decision transmission network	$\{0,1\}$
Second-stage variables:		
$P_{el}^{\omega}[1..H, 1..N]$	Electric output	$\mathbb{R}^+$
$P_{th}^{\omega}[1..H, 1..N]$	Thermal production	$\mathbb{R}^+$
$P_{th,trans}^{\omega}[1..H, 1..H]$	Thermal transmission	$\mathbb{R}$
$P_p^{\omega}[1..H]$	Purchased electricity	$\mathbb{R}^+$
$P_s^{\omega}[1..H]$	Sold electricity	$\mathbb{R}^+$
$s^{\omega}[1..H, 1..N]$	Production state of unit	$\{0,1\}$
$P_{th,netw,abs}^{\omega}[1..H, 1..H]$	Absolute flow on the lines	$\mathbb{R}^+$

5.1.4 Constraints of the Expectation-Based Model

First-stage constraints:

The amount of space is limited: Due to the required space of a CHP unit, we restrict ourselves to investment decisions, leading to at most one CHP unit per house. This is expressed by the constraints

$$\forall s \in S : \sum_{n=1}^{N} Inv\_node[s,n] \leq 1.$$

Only certain houses as potential sites for units: Moreover, not all houses come into question as locations for generating units. The set $\bar{S}$ contains the indices of houses which will not be considered as potential generating nodes in the network. For all $s \in \bar{S}$ the investment decision indicator is set to zero for any kind of unit.

$$\forall i \in \bar{S}, \ \forall n \in \{1,\ldots,N\} : Inv\_node[i,n] = 0.$$

Only certain potential transmission lines: Likewise the investment decision regarding the heat distribution network is limited according to

$$\forall (i,j) \in \bar{E} : Inv\_edge[i,j] = 0. \tag{5.1}$$

Second-stage constraints:

Flow conservation: For every house, the provided heat has to be greater than or equal to the demand. The following inequalities model the energy balance restrictions. For every node i, the produced thermal energy plus the imported energy from neighboring nodes j minus exported energy to neighboring nodes j has to be greater than or equal to the requested thermal energy in node i. Note that import and export does not correspond to the first and second sum, but to the sign of $P^{\omega}_{th,trans}[n,m]$.

$$\forall i \in \{1,\ldots,H\} :$$

$$\sum_{n=1}^{N} P^{\omega}_{th}[i,n] + \sum_{(j,i)\in E} P^{\omega}_{th,trans}[j,i] - \sum_{(i,j)\in E} P^{\omega}_{th,trans}[i,j] \geq D^{\omega}_{th}[i].$$

We assume, that if (n,m) is in E, then (m,n) is not in E. It holds:

$$P^{\omega}_{th,trans}[n,m] \begin{cases} > 0 \Leftrightarrow \text{flow from } n \text{ to } m \\ < 0 \Leftrightarrow \text{flow from } m \text{ to } n. \end{cases}$$

Transmission is bounded: In virtue of the following two inequalities, the variable $P^{\omega}_{th,netw,abs}[i,j]$ is always greater than or equal to $|P^{\omega}_{th,trans}[i,j]|$:

$$P^{\omega}_{th,netw,abs}[i,j] \geq P^{\omega}_{th,trans}[i,j]$$
$$P^{\omega}_{th,netw,abs}[i,j] \geq -P^{\omega}_{th,trans}[i,j].$$

The absolute flow on an edge $(i,j) \in E$ of the network is bounded by a quantity $P^{\omega}_{th,netw,abs,ub}[i,j]$.

$$\forall (i,j) \in E : P^{\omega}_{th,netw,abs,ub}[i,j] \cdot Inv\_edge[i,j] \geq P^{\omega}_{th,netw,abs}[i,j].$$

Supply electric demand: Similar to the thermal demand, the electric demand has to be met in each node. In generating nodes, produced electric energy can be consumed on site, or it can be fed into the main grid. A local distribution is not intended. Shortages of electric energy in generating nodes can be covered by infeeds from the main grid. For the other nodes, using the main grid is the only option to supply electric demand.

$$\forall i \in \{1,\ldots,H\} \; : \; P_p^{\omega}[i] - P_s^{\omega}[i] + \sum_{n=1}^{N} P_{el}^{\omega}[i,n] \geq D_{el}^{\omega}[i].$$

Some technical constraints: The CHP units considered produce electricity and heat in a fixed ratio $a[n]$. Typically the power output of CHP units is not only bounded from above, but also from below as soon as a unit is switched on. We model this aspect by introducing scenario dependent binary variables $s^{\omega}[i,n]$, providing the information whether unit n in house i is switched on or off in scenario ω and the following set of constraints.

$$\forall i \in \{1,\ldots,H\}, \; \forall n \in \{1,\ldots,N\}, \; \forall \omega \in \{1,\ldots,S\} \; :$$
$$Inv\_node[i,n] \geq s^{\omega}[i,n]$$
$$s^{\omega}[i,n] \cdot P_{minmax}[n,1] \leq P_{el}^{\omega}[i,n] \leq s^{\omega}[i,n] \cdot P_{minmax}[n,2]$$
$$P_{th}^{\omega}[i,n] = a[n] \cdot P_{el}^{\omega}[i,n] + s^{\omega}[i,n] \cdot (P_{th,minmax}[n,1] - a[n] \cdot P_{minmax}[n,1]).$$

The first set of constraints models, that a unit has to be purchased in the first stage, before it can be used in the second stage. The second set of constraints models the electric power spectrum of the units, while the third set couples the electric output and the thermal output, which as the electric output is either zero or lies in some interval $[P_{th,minmax}[n,1], P_{th,minmax}[n,2]]$ with $P_{th,minmax}[n,1] > 0$.

5.1.5 Objective Function of the Expected Value Problem

As already mentioned, the investment costs have been annualized.

$$Inv\_costs[n] = \frac{T}{8760} \cdot price[n] \cdot \frac{rate[n] \cdot (rate[n]+1)^{lifetime[n]}}{(rate[n]+1)^{lifetime[n]} - 1}$$

$$Inv\_costs\_netw = \frac{T}{8760} \cdot price\_netw \cdot \frac{rate\_netw \cdot (rate\_netw+1)^{lifetime\_netw}}{(rate\_netw+1)^{lifetime\_netw} - 1}$$

The investment costs for the network are computed per meter. That is the reason for the coefficient $dist[i,j]$ of $Inv\_edge[i,j]$ in the objective.

First-stage costs ($c^T x$):

Annualized investment costs: $\quad \sum_{s \in S} \sum_{n=1}^{N} Inv\_costs[n] \cdot Inv\_node[s,n]$

$\quad + \quad \sum_{(i,j) \in E} Inv\_costs\_netw \cdot dist[i,j] \cdot Inv\_edge[i,j]$

Scenario dependent second stage costs ($q^T y^{\omega}$):

Fuel Costs: $\quad \sum_{s \in S} \sum_{n=1}^{N} C \cdot T \cdot 1/\eta_{el}[n] \cdot P_{el}^{\omega}[s,n]$

Network Costs: $\quad + \quad \sum_{(i,j) \in E} 10^{-6} \cdot T \cdot dist[i,j] \cdot P_{th,netw,abs}^{\omega}[i,j]$

Cost purch. elec.: $\quad + \quad \sum_{i=1}^{H} C_p \cdot T \cdot P_p^{\omega}[s]$

Inc. sold elec.: $\quad + \quad \sum_{s \in S} C_s \cdot T \cdot P_s^{\omega}[s]$

O&M-Cost: $\quad + \quad \sum_{s \in S} \sum_{n=1}^{N} C_{O\&M} \cdot T \cdot P_{el}^{\omega}[s,n]$

5.1.6 Dominance Constrained Model: Risk Averse Preference of Certain Technologies

For any reason we want to built up as many units of a certain type as possible. This could, for example, be due to higher flexibility or reliability, lower environmental pollution, tax advantages or personal preference and so on. That is we want to come to a feasible first-stage decision with a maximal sum of the corresponding binary variables. To avoid critical cost levels, we want to restrict ourselves to those first-stage decisions x leading to random variables $\tilde{f}_x$ which are not worse than a given cost-profile with respect to the increasing convex order.

5.1.7 Results

In our case study, we used the potential topology from Figure 5.1. Furthermore, we considered two types of generating units.

Table 5.3 summarizes some of our computations for the described model. We used a Linux PC with a 3.2GHz Pentium processor and 1GB RAM. The time limit was set to 12 hours. The first column lists the number of benchmark scenarios. The decreasing scaling factor in the second column corresponds to a shifting of the used benchmark distribution to the left. The remaining columns list the number of data scenarios and upper and lower bounds obtained, when applying CPLEX ([60]) and our implementation ddsip.vSD ([54]) of Algorithm 1 as well as the times needed for the computations. As can be seen from Table 5.3, ddsip.vSD was able to solve all instances to optimality or to prove infeasibility (lower bound ∞), respectively, within the horizon of 12 hours. CPLEX, already needing significantly more time to solve the *"2 benchmark scenarios instances"* than ddsip.vSD was in most cases not able to find feasible points for the *"5 benchmark scenarios instances"* within the given horizon of 12 hours. Also the lower bounds are sometimes poor in these cases. CPLEX and ddsip.vSD show the tendency to need more time for lower scaling factors. In some cases ddsip.vSD is able to detect infeasibility very quickly (look up the infeasible instances from Table 5.3). When this happens, then often the reason is that

$$a_{k^*}^* < \max_{\ell=1,\ldots,L} \left\{ \min \left\{ c^\top x + q^\top y : Tx + Wy = z_\ell, \; x \in X, \; y \in Y \right\} \right\}, \qquad (5.2)$$

where k^* denotes the index of the largest outcome of a. The reason for this is the algorithmic realization of Remark 2.18. Then, if (5.2) holds, only a single lower bounding loop has to be passed. The algorithm can be terminated in node one of the outer branch and bound tree as soon as one subproblem is infeasible[20] due to

$$c^\top x + q^\top y > a_{k^*}^*, \; \forall \, (x,y) \in X \times Y : Tx + Wy = z_\ell. \qquad (5.3)$$

During our numerical tests, we could not observe that CPLEX profits from the information $v_{\ell k^*} = 0$.

Table 5.3: Results for network model instances (t.l.e. $\widehat{=}$ time limit of 12 hours was exceeded)

Number of Bench-mark Scenarios	Scaling factor	Number of Data Scenarios	ILOG CPLEX			ddsip.vSD		
			Upper Bound	Lower Bound	Time (sec)	Upper Bound	Lower Bound	Time (sec)
2	0.9	10	-4	-4	1817	-4	-4	1
	0.88	10	-3	-3	3528	-3	-3	22
	0.85	10	-2	-2	762	-2	-2	160
	0.8425	10	-1	-1	14486	-1	-1	2306
	0.97	20	-4	-4	80	-4	-4	1
	0.95	20	-3	-3	227	-3	-3	7
	0.945	20	-2	-2	1700	-2	-2	43
	0.94	20	∞	∞	9253	∞	∞	4
	0.98	30	-4	-4	1476	-4	-4	11
	0.97	30	-3	-3	1887	-3	-3	42
	0.9675	30	-2	-2	7622	-2	-2	62
	0.965	30	-2	-2	3883	-2	-2	53

To be continued on the next page

---

[20]other matters for infeasibility of a subproblem (in node 1) are thinkable, but then even the classical expectation-based problem would be infeasible.

Number of Bench-mark Scenarios	Scaling factor	Number of Data Scenarios	ILOG CPLEX			ddsip.vSD		
			Upper Bound	Lower Bound	Time (*sec*)	Upper Bound	Lower Bound	Time (*sec*)
5	1.15	250	-4	-4	65	-4	-4	5
	1.1	250	-4	-4	52	-4	-4	4
	1.05	250	–	-4	t.l.e.	-3	-3	8662
	1	250	–	-3.65	t.l.e.	∞	∞	183
	1.15	1000	–	-3.65	t.l.e.	-4	-4	245
	1.1	1000	–	-3.65	t.l.e.	-4	-4	383
	1.05	1000	–	-3.65	t.l.e.	-1	-1	28406
	1	1000	–	-3.54	t.l.e.	∞	∞	5.4
	1.2	5000	-4	-4	19061	-4	-4	969
	1.15	5000	-4	-4	17644	-4	-4	1363
	1.135	5000	–	-4	t.l.e.	-4	-4	2107
	1.13	5000	–	-4	t.l.e.	-4	-4	4305
	1.1275	5000	-3	-4	t.l.e.	-4	-4	3956
	1.126	5000	–	-4	t.l.e.	-4	-4	2866
	1.1255	5000	-3	-4	t.l.e.	-4	-4	21018
	1.1	5000	–	-4	t.l.e.	∞	∞	36

Table 5.4 compiles information on problem sizes for the *"5 benchmark scenarios instances"*. Exemplarily we now take a closer look at the first four instances

Table 5.4: Dimensions of mixed-integer linear programming equivalents (for $K = 5$ and some L) and of a single scenario subproblem as it occurs in the lower bounding procedure of our algorithm ($K = 5$).

Number of	subprob. size	250 scen.	1000 scen.	5000 scen.
boolean variables	285	7755	30255	150255
continuous variables	134	33500	134000	670000
constraints	385	33507	133257	665257

with 2 benchmark and 10 data scenarios from Table 5.3. Lowering the scaling factor makes it harder to perform better than the given profile in terms of the increasing convex order. This can also be seen from the increasing objective value (see also the other instances). The investment in certain—at first sight more expensive generating units—is opponent to minimizing expected costs. Figure 5.2 illustrates optimal first-stage decisions regarding the first and the fourth instance from Table 5.3. A circled number in the upper right corner of a house means that there a CHP unit is installed in the first stage of the first instance. A circled number in the lower left corner means that a CHP unit is installed in the first stage of the fourth instance. Our objective was to maximize investments in units of type 2. In the first-stage solution of the expected value problem, all solid lines also occur. In addition the edge between node 4 and node 15 is installed. Regarding the CHP units while minimizing expected costs, it is optimal to install units of type 1 in node 2 and node 14 and units of type 2 in node 6 and node 8.

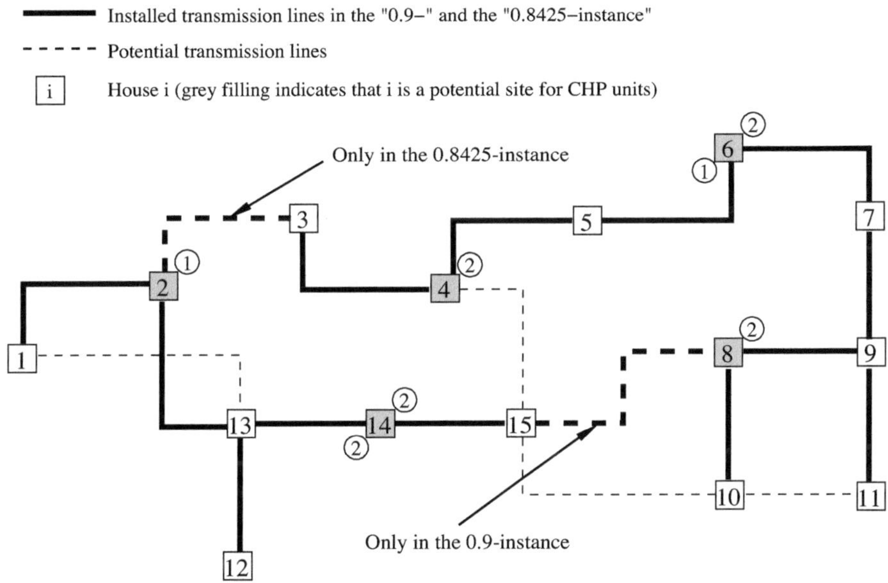

Figure 5.2: First-stage solution of the first and the fourth instance from Table 5.3.

The thin black lines in Figure 5.3 and Figure 5.4 display the graph of the distribution function of to the random variable stemming from Section 5.1.5 with optimal x and optimal y^{ω} $\forall \omega$. The bold black lines in these figures belong to the graphs of the benchmark distributions used in the instances under consideration. The dotted lines represent the resulting cost distributions when using these benchmarks.

Note that the scenario cost distribution resulting from the expected value problem does not fulfill the icx-condition when using the benchmark from Figure 5.4. This can easily be seen, since the costliest scenario exceeds the highest benchmark value.

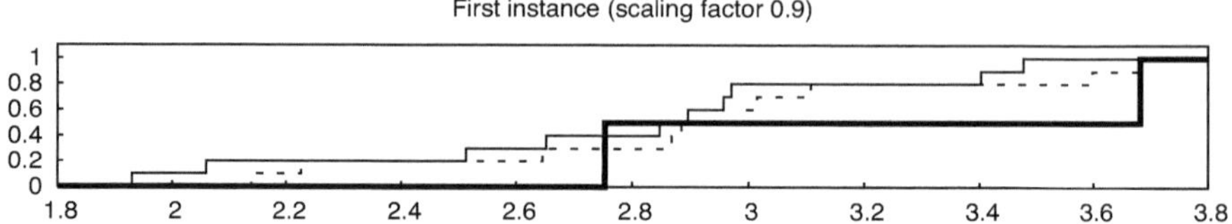

Figure 5.3: Distribution functions belonging to the optimal first-stage decisions of the expectation-based approach (thin black line) and to the model with icx-constraints (dotted line). The bold black line reflects the underlying benchmark profile.

Figure 5.4: The benchmark distribution is successively shifted to the left with a decreasing scaling factor. This makes it harder to perform better than it.

Figure 5.5 shows the integrated survival functions mentioned in Chapters 1 and 2 belonging to the distributions from Figures 5.3 and 5.4. The lower of the two black graphs in Figure 5.5 belongs to the benchmark distribution in Figure 5.4. The dotted lines correspond to the performance functions belonging to the dotted distributions in Figure 5.3 and Figure 5.4 respectively.

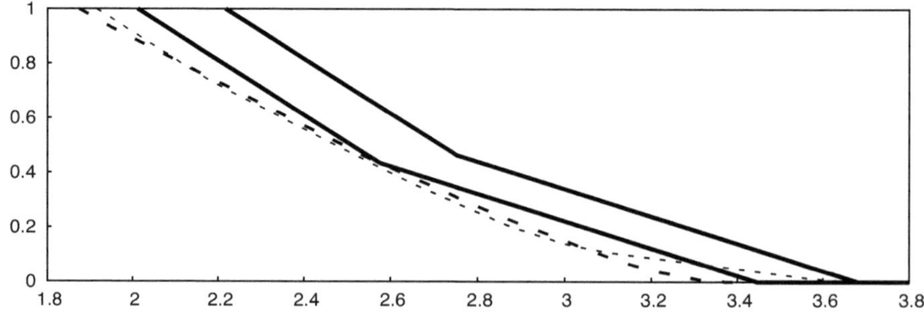

Figure 5.5: Integrated survival functions.

5.2 Sudoku Instances

Sudoku is a popular logic game, which is played over a 9×9 grid, canonically divided into nine 3×3 sub grids. Sudoku begins with some of the grid cells already filled with numbers. The task of a Sudoku player is to fill the remaining empty cells with numbers between 1 and 9 (one number only in each cell), such that each number occurs only once in each row, each column and each of the nine sub blocks. The Sudoku rules can easily be represented with 729 Boolean variables and a system of linear inequalities (cf. [64]).

A two-stage random integer linear program (1.2) arises in the following way: The entries on the main diagonal are chosen as first-stage decisions. Scenarios are formed by single Sudoku puzzles with a small number of prescribed entries and the property that a solution with joint elements on the main diagonal exists. The objective is to minimize the sum of the elements of the secondary diagonal (north-east to south-west).

To arrive at an increasing convex order constrained model (2.1), we choose the objective $g(x) = g^\top x$ as the sum of the elements on the main diagonal. Benchmark scenarios were derived by clustering $\tilde{f}_{x^*}$, where x^* denotes an optimal solution to the expectation model (1.9). In this way we minimize the sum of the main diagonal elements such that the corresponding member of $(\tilde{f}_x)_{x \in X}$ is not worse than the benchmark in terms of the increasing convex order.

We report results with $K = 1$ up to 5 benchmark scenarios and $L = 10$ up to 100 scenarios. Deterministic equivalents according to Proposition 2.16 again become pretty large-scale. Table 5.5 shows dimensions for $K = 5$ and some L.

Table 5.5: Dimensions of mixed-integer linear programming equivalents and of a single scenario subproblem as it occurs in the lower bounding procedure of our algorithm.

Number of	subpr. size	10 scen.	20 scen.	50 scen.	100 scen.
boolean variables	729	7290	14580	36450	72900
general integer variables	9	9	9	9	9
continuous variables	5	50	100	250	500
constraints	743	4195	8385	20955	41905

Table 5.6 summarizes our computations for the Sudoku instances. Again, a Linux-PC with a 3.2GHz Pentium processor and 2GB ram was used. The time limit was set to ten hours.

As before, the first two columns list the numbers K of benchmark and L of data scenarios. The remaining columns list lower and upper bounds obtained when applying CPLEX [60] and our implementation ddsip.vSD of Algorithm 1. Time entries deviating from the limit of 10h indicate that the instance was solved to optimality within this span. It can be seen that ddsip.vSD was able to solve all instances to optimality within the horizon of 10 hours, while CPLEX did not find a feasible point in many cases.

Table 5.6: Results for Sudoku instances

Number of Bench-mark Scenarios	Number of Data Scenarios	CPLEX			ddsip.vSD		
		Upper Bound	Lower Bound	Time (*min*)	Upper Bound	Lower Bound	Time (*min*)
1	10	-	45	600	45	45	4.6
	20	45	45	0.1	45	45	2.8
	50	45	45	0.4	45	45	3
	100	-	45	600	45	45	5.5
2	10	45	45	0.1	45	45	4.6
	20	45	45	0.2	45	45	2.8
	50	45	45	0.9	45	45	3
	100	-	45	600	45	45	5.6
3	10	45	45	0.1	45	45	4.7
	20	45	45	0.3	45	45	2.8
	50	-	45	600	45	45	3
	100	-	45	600	45	45	5.6
4	10	45	45	0.1	45	45	4.6
	20	-	45	600	45	45	2.8
	50	-	45	600	45	45	3
	100	-	45	600	45	45	5.6
5	10	45	45	0.2	45	45	4.6
	20	45	45	332	45	45	2.8
	50	-	45	600	45	45	3
	100	-	45	600	45	45	5.6

In our paper [50], we reported preliminary results for the Sudoku instances. Since then we were able to speed up our algorithm. The main reason for the significant improvement of computing times is the implementation of Corollary 2.16. In [50] we came from Proposition 2.13.

Chapter 6

An Alternative Formulation for Optimization under Stochastic Dominance Constraints

In [74], a novel formulation for first- and second order stochastic dominance is introduced. In this chapter we discuss relevant conclusions for our framework and advanced algorithmic perspectives. The theory in [74] is developed in a maximization setting, where larger outcomes of the random variables are preferred to smaller outcomes. The results presented in [74] imply the following proposition for our specific random variables originating from mixed-integer value functions. In Proposition 6.1 we start out from a maximization setting before we switch to the preference of smaller outcomes.

Proposition 6.1. *Let z and a follow discrete distributions with only finitely many realizations z_ℓ, $\ell = 1,\ldots,L$, and $a_1 < \ldots < a_K$ as well as probabilities π_ℓ, $\ell = 1,\ldots,L$, and p_k, $k = 1,\ldots,K$, respectively. Let further g be linear. Assume (A1) and (A2). Then $\max\{g^\top x : \tilde{h}_x \succeq_2 a, x \in X\}$ is equivalent to*

$$
\max\left\{ g^\top x : \begin{array}{lll}
\tilde{h}_x(z^{-1}(z_\ell)) & \geq \sum_{j=1}^{K} a_j v_{\ell j} & \forall \ell \\
\sum_{\ell=1}^{L} \pi_\ell \sum_{j=1}^{k-1}(a_k - a_j)v_{\ell j} & \leq \mathbb{E}[(a_k - a)_+] & \forall k \\
\sum_{j=1}^{K} v_{\ell j} & = 1 & \forall \ell \\
x \in X,\ v_{\ell k} \geq 0 & & \forall \ell \quad \forall k
\end{array} \right\} \tag{6.1}
$$

Recall that $\tilde{h}_x(.) := c^\top x + \max_{y \in Y}\{q^\top y : Tx + Wy = z(.)\}$. That is, to end up with a mixed-integer linear program, we have to ensure, that the left-hand side of the first set of constraints *can* attain the value of $\tilde{h}_x$ in the correspondent scenario. Since $\tilde{h}_x(z^{-1}(z_\ell)) \in \{c^\top x + q^\top y : Tx + Wy = z_\ell, y \in Y\}$ this can be achieved by replacing $\tilde{h}_x(z^{-1}(z_\ell))$ by $c^\top x + q^\top y_\ell$ and the additional constraints $Tx + Wy_\ell = z_\ell$ and $y_\ell \in Y$ for all $\ell = 1,\ldots,L$.

Proposition 6.2. *In a framework, where smaller outcomes of the random variables are preferred to larger outcomes, (6.1) under consideration of the last passage turns into*

$$\min \left\{ g^\top x : \quad
\begin{array}{lll}
c^\top x + q^\top y_\ell & \leq \sum_{j=1}^{K} a_j v_{\ell j} & \forall \ell \\
Tx + W y_\ell & = z_\ell & \forall \ell \\
\sum_{\ell=1}^{L} \pi_\ell \sum_{j=k+1}^{K} (a_j - a_k) v_{\ell j} & \leq \mathbb{E}[(a - a_k)_+] & \forall k \\
\sum_{j=1}^{K} v_{\ell j} & = 1 & \forall \ell \\
x \in X, \; y_\ell \in Y, \; v_{\ell k} & \geq 0 & \forall \ell \;\; \forall k
\end{array}
\right\} \quad (6.2)$$

Proof. Since f is convex iff $-f$ is concave, it is clear that Definitions 1.6 and 1.2 yield $\tilde{f}_x \leq_{icx} d \Leftrightarrow -\tilde{f}_x \succeq_2 -d$ and thus:

$$\begin{aligned}
& \min \left\{ \; g(x) : \quad \tilde{f}_x \leq_{icx} a \,, \, x \in X \right\} \\
= \; & -\max \left\{ -g(x) : \; -\tilde{f}_x \succeq_2 -a \,, \, x \in X \right\}.
\end{aligned}$$

Let $\tilde{a}_1 < \ldots < \tilde{a}_K$ denote the realizations of $-a$. We refer to the realizations of a as $-\tilde{a}_K =: a_1 < \ldots < a_K := -\tilde{a}_1$. According to 6.1 the last expression is equal to

$$-\max \left\{ -g^\top x : \quad
\begin{array}{ll}
-\tilde{f}_x(z^{-1}(z_\ell)) & \geq \sum_{j=1}^{K} \tilde{a}_j \tilde{v}_{\ell j} \\
\sum_{\ell=1}^{L} \pi_\ell \sum_{j=1}^{k-1} (\tilde{a}_k - \tilde{a}_j) \tilde{v}_{\ell j} & \leq \mathbb{E}[(\tilde{a}_k - (-a))_+] \\
\sum_{j=1}^{K} \tilde{v}_{\ell j} & = 1 \\
x \in X, \; \tilde{v}_{\ell k} \geq 0 &
\end{array}
\right\}$$

(For ease of presentation, we left out the $\ell = 1, \ldots, L$ and $k = 1, \ldots, K$ statements. In the next step we express $\tilde{a}_k$ in terms of realizations of a.)

$$= \min \left\{ g^\top x : \quad
\begin{array}{ll}
-\tilde{f}_x(z^{-1}(z_\ell)) & \geq -\sum_{j=1}^{K} a_{K-j+1} \tilde{v}_{\ell j} \\
\sum_{\ell=1}^{L} \pi_\ell \sum_{j=1}^{k-1} (a_{K-j+1} - a_{K-k+1}) \tilde{v}_{\ell j} & \leq \mathbb{E}[(a - a_{K-k+1})_+] \\
\sum_{j=1}^{K} \tilde{v}_{\ell j} & = 1 \\
x \in X, \; \tilde{v}_{\ell k} \geq 0 &
\end{array}
\right\}$$

(Next we invert the order of summation in the first and second set of constraints. In the second set of constraints we also alter the order in which they occur in the problem. The former k-th constraint in the second set of constraints becomes the $(K - k + 1)$-th one.)

$$= \min \left\{ g^\top x : \quad
\begin{array}{ll}
\tilde{f}_x(z^{-1}(z_\ell)) & \leq \sum_{j=1}^{K} a_j \tilde{v}_{\ell(K-j+1)} \\
\sum_{\ell=1}^{L} \pi_\ell \sum_{j=k+1}^{K} (a_j - a_k) \tilde{v}_{\ell(K-j+1)} & \leq \mathbb{E}[(a - a_k)_+] \\
\sum_{j=1}^{K} \tilde{v}_{\ell j} & = 1 \\
x \in X, \; \tilde{v}_{\ell k} \geq 0 &
\end{array}
\right\}$$

$$= \min \left\{ g^\top x : \begin{array}{ll} \tilde{f}_x(z^{-1}(z_\ell)) & \leq \sum_{j=1}^{K} a_j v_{\ell j} \\ \sum_{\ell=1}^{L} \pi_\ell \sum_{j=k+1}^{K} (a_j - a_k) v_{\ell j} & \leq \mathbb{E}[(a - a_k)_+] \\ \sum_{j=1}^{K} v_{\ell j} & = 1 \\ x \in X, \ v_{\ell k} \geq 0 \end{array} \right\}$$

$$= \min \left\{ g^\top x : \begin{array}{ll} c^\top x + q^\top y_\ell & \leq \sum_{j=1}^{K} a_j v_{\ell j} \\ Tx + Wy_\ell & = z_\ell \\ \sum_{\ell=1}^{L} \pi_\ell \sum_{j=k+1}^{K} (a_j - a_k) v_{\ell j} & \leq \mathbb{E}[(a - a_k)_+] \\ \sum_{j=1}^{K} v_{\ell j} & = 1 \\ x \in X, \ y_\ell \in Y, \ v_{\ell k} \geq 0 \end{array} \right\} \qquad (6.3)$$

$\square$

Regarding lower bounds, the decomposition structure of problem (6.2) is more or less identical to the one we already got to know. In analogy to the previous discussion the coupling dominance modeling constraints induce the Lagrangean

$$\mathcal{L}(x_1, \ldots, x_L, \Delta, \lambda) = \sum_{\ell=1}^{L} \pi_\ell \mathcal{L}_\ell(x_\ell, \Delta_{\ell \bullet}, \lambda)$$

with

$$\mathcal{L}_\ell(x_\ell, \Delta_{\ell \bullet}, \lambda) := g^\top x_\ell + \sum_{k=1}^{K} \lambda_k \left[\sum_{j=k+1}^{K} (a_j - a_k) v_{\ell j} - \mathbb{E}\left[(a - a_k)_+ \right] \right].$$

Unfortunately, the decomposition effect with respect to upper bounds (cf. Algorithm 2, line 5) breaks down. Customizing Algorithm 2 to qualify it for the particularities of the new formulation gives rise to the following scenario subproblems ($\ell = 1, \ldots, L$).

$$\min \left\{ \sum_{k=1}^{K} \sum_{j=k+1}^{K} (a_j - a_k) v_{\ell k} : \begin{array}{ll} c^\top \bar{x} + q^\top y_\ell & \leq \sum_{j=1}^{K} a_j v_{\ell j} \\ T\bar{x} + Wy_\ell & = z_\ell \\ \sum_{k=1}^{K} v_{\ell k} & = 1 \\ y_\ell \in Y, \ v_{\ell k} \geq 0, & k = 1, \ldots, K \end{array} \right\}$$

Because of constraints coupling different benchmark scenarios, the sum over k cannot be pulled out of these problems (cf. (4.5)). Therefore it is not guaranteed, that the choice for the $v_{\ell k}$ minimizing the sum over k also minimizes the weighted sum over ℓ below, which is fundamental for the feasibility of $\bar{x}$ (see line 9 in Algorithm 2).

$$\sum_{\ell=1}^{L} \pi_\ell \underbrace{\sum_{j=k+1}^{K} (a_j - a_k) v_{\ell j}}_{(*)} \leq \mathbb{E}[(a - a_k)_+] \qquad \forall k = 1, \ldots, K$$

The crucial point is that $(*)$ might be smaller for a different choice of the $v_{\ell k}$. As an undesired consequence it can happen that $\bar{x}$ is neglected due to awkwardly selected $v_{\ell k}$ even though the $\bar{x}$ from the heuristic was optimal. A way out, possibly leading to long computations would be to fix x to $\bar{x}$ in (6.2) and solve this large problem to check the feasibility of the suggestion $\bar{x}$. On the other hand, computations on this problem "only" consist of finding a feasible second stage, while the subproblems in Algorithm 2 have to be solved to optimality. This idea of course is also an alternative to Algorithm 2 while dealing with the previous problem formulation. However, which procedure is favorable likely depends on the concrete problem of interest. Note that the information on $\tilde{f}_x$'s distribution is lost for similar reasons as discussed in Section 2.2.

In the following we compare results obtained from utilizing the algorithm from Chapter 4 and from applying the ideas just mentioned. As a test instance we again consider a representative ($L = 1000$ data scenarios and $K = 5$ benchmark scenarios) of the investment planning problem described on pages 46–46 (see also [52, 73]). Before we discuss the performance of the different decomposition methods, we present some findings obtained from applying CPLEX to deterministic equivalents. As expected CPLEX is able to solve the deterministic equivalents very efficiently (cf. Table 3.3). To eliminate measurement uncertainty we solved the deterministic equivalents 10 times each. The results are compiled in Table 6.1. In the row *Sol. time*, μ and σ represent the average computing time and the standard deviation of the calculating times in seconds, respectively.

In both formulations, the number of variables results from the number of second-stage variables (including the v_j-variables) times the number of scenarios plus the number of first-stage variables ($101 \cdot 1000 + 4$). The total number of constraints arises from 2 constraints constituting X, plus the number of scenarios times the number of second-stage constraints, which is 28. In the Luedtke-formulation adding $2L + (K - 1)$ to this number ($\mathcal{O}(L + K)$ constraints) yields the total number of constraints (cf. (6.2)), while in our formulation we have $\mathcal{O}(L \cdot K)^{21}$ constraints

---

[21] exactly $(L + 1) \cdot K$

Table 6.1: Results from solving deterministic equivalents

	Luedtke's formulation	Our formulation
Sol. time	$\mu = 23.082$ ($\sigma = 0.1685$)	$\mu = 35.6430$ ($\sigma = 0.0823$)
Variables	Nneg: 101000, General Int.: 4	Nneg: 101000, General Int.: 4
Constraints	$\leq 5005, \geq 24001, = 1000$, tot.: 30006	$\leq 9006, \geq 24001$, tot.: 33007
Nonzeros	316008	706008
Iterations	66530	49672
Nodes	0 (0 cuts, solved by heuristic)	0 (0 cuts, solved by heuristic)
Obj. value	188	188

in addition (cf. (2.13)). As can be seen from Table 6.1, Luedtke's formulation introduces considerably less nonzeros, which is computationally beneficial. The sparser constraint matrix in Luedtke's formulation is surely one reason for the lesser average time per iteration (approximately factor 2).

Figure 6.1 illustrates how upper and lower bounds evolve over the number of nodes in the outer branch and bound trees, when using decomposed models. The horizontal dashed line at 188 represents the "development" of upper bounds, i. e., the heuristic identifies the optimum already in the first node of the tree. Regarding lower bounds, it can be seen that our formulation yields better bounds for smaller trees. After 800 nodes the lower bound is set to 187 when using our problem formulation and to 186, when using Luedtke's formulation. An important point now is, how long it takes to compute 800 nodes using the different decomposition schemes. This information can be found in figure 6.2. It is striking that between node 100 and node 200 the lower bounding is more time consuming using Luedtke's formulation than our formulation, while the graphs are more or less parallel for the rest of the nodes. After 800 nodes, using lower bounding subproblems stemming from Luedtke's formulation led to a 45 minutes longer computation time (only lower bounds) compared to the use of our formulation. "Our" value of the lower bound was also slightly better after 800 nodes: 187 versus 186 (these gaps did not vanish after more than 2800 nodes and more than 15 hours of computing time). Regarding upper bounds, fixing first-stage suggestions in the deterministic equivalent as described above, systematically led to longer computation times than the decomposition with respect to upper bounds from Chapter 4. The distance between the two graphs becomes larger with an increasing number of nodes. After 800 nodes the difference added up to 68 minutes.

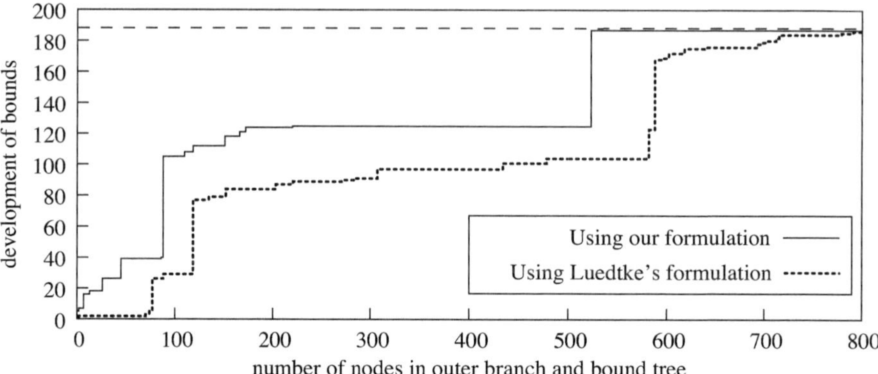

Figure 6.1: Comparison of the development of upper and lower bounds using the different formulations during the decomposition procedure

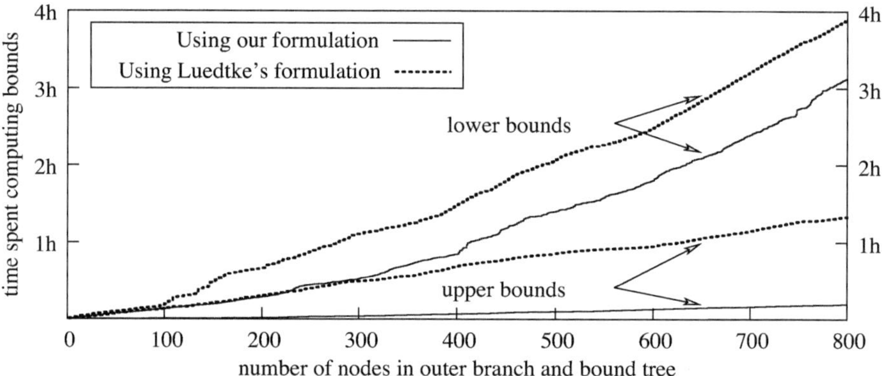

Figure 6.2: Comparison of the times needed to compute upper and lower bounds using the different formulations during the decomposition procedure

References

[1] C. ACERBI AND D. TASCHE, *On the coherence of expected shortfall*, Journal of Banking and Finance **26** (2002), pp. 1487–1503. [5]

[2] S. AHMED, *Convexity and decomposition of mean-risk stochastic programs*, Mathematical Programming **106** (2006), pp. 433–446. [5]

[3] S. ALBERS, *Online algorithms*, Mathematical Programming **97** (2003), pp. 3–26. [1]

[4] C. D. ALIPRANTIS AND K. C. BORDER, *Infinite-Dimensional Analysis*, Springer, Berlin, 1999. [31]

[5] R. ANBIL AND F. BARAHONA, *On some difficult linear programs coming from set partitioning*, Discrete Applied Mathematics **118** (2002), pp. 3–11. [53]

[6] J. M. ARROYO, M. CARRIÓN, AND A. J. CONEJO, *Forward contracting and selling price determination for a retailer*, IEEE Transactions on Power Systems **22** (2007), pp. 2105–2114. [33, 36, 40]

[7] J. M. ARROYO, M. CARRIÓN, A. J. CONEJO, AND A. B. PHILPOTT, *A stochastic programming approach to electric energy procurement for large consumers*, IEEE Transactions on Power Systems **22** (2007), pp. 744–754. [33]

[8] P. ARTZNER, F. DELBAEN, J. M. EBER, AND D. HEATH, *Coherent measures of risk*, Mathematical Finance **9** (1999), pp. 203–228. [5]

[9] S. BALAKRISHNAN, A. J. CONEJO, S. A. GABRIEL, AND M. A. PLAZAS, *Optimal price and quantity determination for retail electric power contracts*, IEEE Transactions on Power Systems **21** (2006), pp. 180–187. [33]

[10] B. BANK, J. GUDDAT, D. KLATTE, B. KUMMER, AND K. TAMMER, *Nonlinear Parametric Optimization*, Akademie-Verlag, Berlin, 1982. [20]

[11] B. BANK AND R. MANDEL, *Parametric Integer Optimization*, Akademie-Verlag, Berlin, 1988. [6]

[12] F. BASTIN, C. CIRILLO, AND P. TOINT, *Theory for nonconvex stochastic programming with an application to mixed logit*, Mathematical Programming **108** (2006), pp. 207–234.   [1]

[13] E. M. L. BEALE, *On minimizing a convex function subject to linear inequalities*, Journal of the Royal Statistical Society Series B **17** (1955), pp. 173–184.   [1]

[14] E. M. L. BEALE, J. J. H. FORREST, AND C. J. TAYLOR, *Multi-time-period stochastic programming*, in: Stochastic Programming, Dempster, M. A. H., ed., 1980, pp. 387–402.   [7]

[15] A. BEN-TAL AND A. NEMIROVSKI, *Robust optimization – methodology and applications*, Mathematical Programming **92** (2002), pp. 453–480. [1]

[16] D. BERTSIMAS, *The price of robustness*, Operations Research **52** (2004), pp. 35–53.   [1]

[17] P. BILLINGSLEY, *Convergence of Probability Measures*, Wiley, New York, 1968.   [14, 15, 16, 20]

[18] P. BILLINGSLEY, *Probability and Measure*, Wiley, New York, 1986.   [18, 19]

[19] J. R. BIRGE, *Decomposition and partitioning methods for multistage stochastic linear programs*, Operations Research **33** (1985), pp. 989–1007.   [7]

[20] J. R. BIRGE AND F. LOUVEAUX, *Introduction to Stochastic Programming*, Springer, New York, 1997.   [3, 4, 7, 21, 26, 33, 50]

[21] C. E. BLAIR AND R. G. JEROSLOW, *The value function of a mixed integer program: I*, Discrete Mathematics **19** (1977), pp. 121–138.   [6]

[22] K.-H. BORGWARDT, *Optimierung, Operations Research, Spieltheorie*, Birkhäuser, 2001.   [55]

[23] A. BROOKE, D. KENDRICK, A. MEERAUS, AND R. RAMAN, *GAMS: A User's Guide*, GAMS Development Corporation, Washington, 1998.   [40]

[24] C. C. CARØE AND R. SCHULTZ, *Dual decomposition in stochastic integer programming*, Operations Research Letters **24** (1999), pp. 37–45. [26, 50]

[25] M. CARRIÓN, U. GOTZES, AND R. SCHULTZ, *Risk aversion for an electricity retailer with second-order stochastic dominance constraints*, Computational Management Science **6** (2009), pp. 233–250.   [VII, 2, 33]

[26] F. R. CHANG, *Stochastic optimization in continuous time*, Cambridge University Press, 2004.   [1]

[27] A. J. CONEJO, R. CONTRERAS, R. ESPÍNOLA, AND F. J. NOGALES, *Forecasting next-day electricity prices by time series models*, IEEE Transactions on Power Systems **17** (2002), pp. 342–348.   [36]

[28] A. J. CONEJO, R. CONTRERAS, R. ESPÍNOLA, AND F. J. NOGALES, *Arima models to predict next-day electricity prices*, IEEE Transactions on Power Systems **18** (2003), pp. 1014–1020.   [36]

[29] A. J. CONEJO, R. CONTRERAS, R. ESPÍNOLA, AND M. A. PLAZAS, *Forecasting electricity prices for a day-ahead pool-based electric energy market*, International Journal of Forecasting **21** (2005), pp. 435–462.   [36]

[30] A. J. CONEJO, M. A. PLAZAS, AND F. J. PRIETO, *Multimarket optimal bidding for a power producer*, IEEE Transactions on Power Systems **20** (2005), pp. 2041–2050.   [33]

[31] S. CONTI, H. HELD, M. PACH, M. RUMPF, AND R. SCHULTZ, *Shape optimization under uncertainty—a stochastic programming perspective*, SIAM Journal on Optimization **19** (2009), pp. 1610–1632.   [1]

[32] G. B. DANTZIG, *Linear programming under uncertainty*, Management Science **1** (1955), pp. 197–206.   [1]

[33] B. A. DAVEY AND H. A. PRIESTLEY, *Introduction to Lattices and Order*, Cambridge University Press, New York, 2002.   [31]

[34] F. DELBAEN, *Coherent risk measures on general probability spaces*, in: Advances in Finance and Stochastics: Essays in Honour of Dieter Sondermann, Sandmann, K. and Schönbucher P. J., eds., 2007, pp. 1–39.   [5]

[35] D. DENTCHEVA, R. HENRION, AND A. RUSZCZYŃSKI, *Stability and sensitivity of optimization problems with first order stochastic dominance constraints*, SIAM Journal on Optimization **18** (2007), pp. 322–337.   [9, 20]

[36] D. DENTCHEVA AND A. RUSZCZYŃSKI, *Stochastic optimization with dominance constraints*, SIAM Journal on Optimization **14** (2003), pp. 548–566.   [9]

82   References

[37] D. DENTCHEVA AND A. RUSZCZYŃSKI, *Optimality and duality theory for stochastic optimization problems with nonlinear dominance constraints*, Mathematical Programming **99** (2004), pp. 329–350. [9]

[38] DEUTSCHER WETTERDIENST, *Klimadaten im KL-Standardformat*, Deutscher Wetterdienst, http://www.dwd.de/. [61]

[39] P. G. L. DIRICHLET, *Verallgemeinerung eines Satzes aus der Lehre von den Kettenbrüchen nebst einigen Anwendungen auf die Theorie der Zahlen*, Bericht über die zur Bekanntmachung geeigneten Verhandlungen der Königlich Preussischen Akademie der Wissenschaften zu Berlin (1842), pp. 93–95. [4]

[40] D. DRAPKIN AND R. SCHULTZ, *An algorithm for stochastic programs with first-order dominance constraints induced by linear recourse*, Discrete Applied Mathematics (to appear). [21, 58]

[41] N. DUNFORD AND J. T. SCHWARTZ, *Linear Operators, Part I: General Theory*, Interscience Publishers, Inc., New York, 1957. [15]

[42] DUPAČOVÁ, *Multistage stochastic programs: The state-of-the-art and selected bibliography*, Kybernetika **31** (1995), pp. 151–174. [7]

[43] A. EICHHORN AND W. RÖMISCH, *Polyhedral risk measures in stochastic programming*, SIAM Journal on Optimization **16** (2005), pp. 69–95. [5, 7]

[44] J. ELSTRODT, *Maß- und Integrationstheorie*, Springer, Berlin, 2002. [8]

[45] J. ELTON AND T. P. HILL, *Fusions of a probability distribution*, Annals of Probability **20** (1992), pp. 421–454. [8]

[46] P. C. FISHBURN, *Mean-risk analysis with risk associated with below-target returns*, American Economic Review **67** (1977), pp. 116–126. [10]

[47] W. H. FLEMING AND R. W. RISHEL, *Deterministic and stochastic optimal control*, Springer, New York, 1975. [1]

[48] S.-E. FLETEN, T. T. LIE, B. K. POKHAREL, AND G. B. SHRESTHA, *Medium term power planning with bilateral contracts*, IEEE Transactions on Power Systems **20** (2005), pp. 627–633. [33]

[49] R. GOLLMER, U. GOTZES, F. NEISE, AND R. SCHULTZ, *Risk modelling via stochastic dominance in power systems with dispersed generation*, Preprint 651–2007 Department of mathematics, University Duisburg-Essen, http://www.uni-due.de/~hn215go/gotzes/preprint651.pdf, accepted for presentation, International Conference on Applications to Power Systems, ISAP, Taiwan, 2007. [2]

[50] R. GOLLMER, U. GOTZES, AND R. SCHULTZ, *Second-order stochastic dominance constraints induced by mixed-integer linear recourse*, Preprint 644–2007 Department of mathematics, University Duisburg-Essen, http://www.uni-due.de/~hn215go/gotzes/preprint644.pdf (2007). [2, 61, 72]

[51] R. GOLLMER, U. GOTZES, AND R. SCHULTZ, *A note on second-order stochastic dominance constraints induced by mixed-integer linear recourse*, Mathematical Programming, http://dx.doi.org/10.1007/s10107-009-0270-0 (to appear). [2]

[52] R. GOLLMER, F. NEISE, AND R. SCHULTZ, *Stochastic programs with first-order dominance constraints induced by mixed-integer linear recourse*, SIAM Journal on Optimization **19** (2008), pp. 552–571. [13, 20, 21, 26, 44, 76]

[53] E. GÓMEZ-VILLALVA AND A. RAMOS, *Optimal energy management of an industrial consumer in liberalized markets*, IEEE Transactions on Power Systems **18** (2003), pp. 716–723. [33]

[54] U. GOTZES AND F. NEISE, *User's guide to ddsip.vSD—A C Package for the Dual Decomposition of Stochastic Programs with Dominance Constraints Induced by Mixed-Integer Linear Recourse*, Universität Duisburg-Essen, http://www.uni-due.de/~hn215go/gotzes/ddsip.vSD-man.pdf, 2007. [2, 54, 66]

[55] U. GOTZES AND R. SCHULTZ, *Risikoaversion mittels stochastischer Dominanz mit Anwendungen bei Optimierungsproblemen in der Energiewirtschaft*, Optimierung in der Energiewirtschaft, VDI-Tagung Leverkusen, 27. – 28. November 2007, Tagungsband, hrsg. v. Verein Deutscher Ingenieure, VDI Verlag, Düsseldorf, VDI-Berichte Band **2018** (2007), pp. 221–235. [2, 61]

[56] U. GOTZES, O. WOLL, R. SCHULTZ, AND C. WEBER, *Verteilte Erzeugung im liberalisierten Energiemarkt – Analyse von Investitionsentschei-*

dungen, in: Modellierung und Optimierung von Energiesystemen, Schultz, R. and Wagner, H.-J., eds., 2008. [61]

[57] T. HEINZE AND R. SCHULTZ, *A branch-and-bound method for multistage stochastic integer programs with risk objectives*, Optimization **57** (2008), pp. 277–293. [7]

[58] M. HELLWIG, *Entwicklung und Anwendung parametrisierter Standard-Lastprofile*, Fakultät für Elektrotechnik und Informationstechnik, Technische Universität München, Dissertation, http://deposit.ddb.de/cgi-bin/dokserv?idn=969627017, 2003. [61]

[59] C. HELMBERG AND K. C. KIWIEL, *A spectral bundle method with bounds*, Mathematical Programming (see also: http://www-user.tu-chemnitz.de/~helmberg/ConicBundle/Manual/) **93** (2002), pp. 173–194. [53]

[60] ILOG, *Ilog cplex 9*, http://www.ilog.com/products/cplex/ (2003). [26, 66, 71]

[61] ILOG, *Ilog cplex 10*, http://www.ilog.com/products/cplex/ (2006). [40]

[62] J. L. W. V. JENSEN, *Sur les fonctions convexes et les inégalités entre les valeurs moyennes*, Acta Mathematica **30** (1906), pp. 175–193. [8]

[63] D. KAHNEMAN AND A. TVERSKY, *Prospect theory: An analysis of decisions under risk*, Econometrica **47** (1979), pp. 313–327. [7]

[64] V. KAIBEL AND T. KOCH, *Mathematik für den Volkssport*, DMV-Mitteilungen **14** (2006), pp. 93–96. [61, 71]

[65] P. KALL AND S. W. WALLACE, *Stochastic Programming*, Wiley, Chichester, 1994. [3, 4, 7, 21, 26, 50]

[66] H. G. KELLERER, *Markov-Komposition und eine Anwendung auf Martingale*, Mathematische Annalen **198** (1972), pp. 99–122. [31]

[67] B. W. KERNIGHAN AND D. M. RITCHIE, *The C Programming Language, Second Edition*, Prentice Hall, Inc., 1988. [2]

[68] R. P. KERTZ AND U. RÖSLER, *Stochastic and convex orders and lattices of probability measures, with a martingale interpretation.*, Israel Journal of Mathematics **77** (1992), pp. 129–164. [31]

[69] O. KLAAR, *Algorithmische Ansätze zur stochastischen Optimierung unter Dominanznebenbedingungen*, Diploma Thesis, Department of Mathematics, University of Duisburg-Essen (2009).   [21]

[70] T. K. KRISTOFFERSON, *Deviation measures in two-stage stochastic linear programming*, Mathematical Methods of Operations Research **62** (2006), pp. 255–274.   [5]

[71] H. LEVY, *Stochastic dominance and expected utility: Survey and analysis*, Management Science **38** (1992), pp. 555–593.   [9]

[72] F. V. LOUVEAUX, *A solution method for multistage stochastic programs with recourse with applications to an energy investment problem*, Operations Research **28** (1980), pp. 889–902.   [7]

[73] F. V. LOUVEAUX AND Y. SMEERS, *Optimal investments for electricity generation: A stochastic model and a test problem*, in: Numerical Techniques for Stochastic Optimization, Ermoliev, Yu. and Wets, R. J.-B., eds., 1988, pp. 445–454.   [46, 76]

[74] J. LUEDTKE, *New formulations for optimization under stochastic dominance constraints*, SIAM Journal on Opzimization **19** (2008), pp. 1433–1450.   [73]

[75] A. MÄRKERT, *Deviation measures in stochastic programming with mixed-integer recourse*, Universität Duisburg-Essen, Campus Duisburg, Dissertation, http://www.ub.uni-duisburg.de/ETD-db/theses/available/ duett-04272004-161939/unrestricted/maerkertdiss.pdf, 2004.   [VII, 5, 10]

[76] A. MÄRKERT, *User's guide to ddsip—A C Package for the Dual Decomposition of Stochastic Programs with Mixed-Integer Linear Recourse*, Universität Duisburg-Essen, http://www.uni-due.de/~hn215go/ software/ddsip-man.pdf, 2004.   [VII]

[77] A. MÄRKERT AND R. SCHULTZ, *On deviation measures in stochastic integer programming*, Operations Research Letters **33** (2005), pp. 441–449.   [5, 7]

[78] A. MÜLLER AND M. SCARSINI, *Stochastic order real and lattices of probability and measure*, SIAM Journal on Optimization **16** (2006), pp. 1024–1043.   [30, 31]

[79] A. MÜLLER AND D. STOYAN, *Comparison Methods for Stochastic Models and Risks*, John Wiley and Sons, Chichester, UK, 2002.   [8, 9, 10, 15]

[80] F. NEISE, *Risk Management in Stochastic Integer Programming: With Application to Dispersed Power Generation*, Vieweg+Teubner, 2008.   [13]

[81] G. L. NEMHAUSER AND L. A. WOLSEY, *Integer and Combinatorial Optimization*, Wiley, New York, 1988.   [52]

[82] N. NOYAN, G. RUDOLF, AND A. RUSZCZYŃSKI, *Relaxations of linear programming problems with first order stochastic dominance constraints*, Operations Research Letters **34** (2006), pp. 653–659.   [13, 26]

[83] W. OGRYCZAK AND A. RUSZCZYŃSKI, *From stochastic dominance to mean-risk models: Semideviations as risk measures*, European Journal of Operations Research **116** (1999), pp. 33–50.   [10, 12]

[84] W. OGRYCZAK AND A. RUSZCZYŃSKI, *On consistency of stochastic dominance and mean-semideviation models*, Mathematical Programming **89** (2001), pp. 217–232.   [10]

[85] W. OGRYCZAK AND A. RUSZCZYŃSKI, *Dual stochastic dominance and related mean-risk models*, SIAM Journal on Optimization **13** (2002), pp. 60–78.   [5]

[86] OMEL, *Market Operator of the Electricity Market of Mainland Spain*, http://www.omel.es.   [39]

[87] OMIP, *Forward and Futures Market of the Iberian Electricity Market, Spain and Portugal*, http://www.omip.pt.   [40]

[88] T. PENNANEN, *Epi-convergent discretizations of multistage stochastic programs*, Mathematics of Operations Research **30** (2005), pp. 245–256.   [1]

[89] C. G. PFLUG AND W. RÖMISCH, *Modeling, Measuring and Managing Risk*, World Scientific Publishing, Singapore, 2007.   [2]

[90] G. C. PFLUG, *Some remarks on the value-at-risk and the conditional value-at-risk*, in: Probabilistic Constrained Optimization: Methodology and Applications, Uryasev, S., ed., 2000, pp. 272–281.   [5]

[91] B. T. POLYAK, *A general method for solving extremum problems*, Soviet Mathematics Doklady **8** (1967), pp. 593–597.   [53]

[92] B. T. POLYAK, *Minimization of unsmooth functionals*, USSR Computational Mathematics and Mathematical Physics **9** (1969), pp. 509–521. [53]

[93] A. PRÉKOPA, *Probabilistic programming*, in: Stochastic Programming, Handbooks of Operations Research and Management Science, Ruszczyński, A. and Shapiro, A., eds., 2003, pp. 2671–345. [7]

[94] A. PRÉKOPA, *Stochastic Programming*, Kluwer, Dordrecht, 1995. [3, 4, 7, 21, 26, 50]

[95] R. T. ROCKAFELLAR AND S. URYASEV, *Optimization of conditional value-at-risk*, Journal of Risk **2** (2000), pp. 21–41. [5]

[96] R. T. ROCKAFELLAR AND S. URYASEV, *Conditional value-at-risk for general loss distributions*, Journal of Banking and Finance **26** (2002), pp. 1443–1471. [5]

[97] W. RÖMISCH, *Stability of stochastic programming problems*, in: Stochastic Programming, Handbooks of Operations Research and Management Science, Ruszczyński, A. and Shapiro, A., eds., 2003, pp. 483–554. [7]

[98] W. RÖMISCH AND R. SCHULTZ, *Multistage stochastic integer programs: an introduction*, in: Online optimization of large scale systems, Grötschel, M.; Krumke S. O. and Rambau, J., eds., 2001, pp. 581–622. [7]

[99] A. RUSZCZYŃSKI, *Decomposition methods in stochastic programming*, Mathematical Programming **79** (1997), pp. 333–353. [26, 50]

[100] A. RUSZCZYŃSKI AND A. SHAPIRO, *Stochastic programming models*, in: Stochastic Programming, Handbooks of Operations Research and Management Science, Ruszczyński, A. and Shapiro, A., eds., 2003, pp. 1–64. [1, 2, 3, 4, 21, 26, 50]

[101] A. SCHRIJVER, *Theory of linear and integer programming*, Wiley–Interscience, 1999. [4, 52]

[102] R. SCHULTZ, *Some aspects of stability in stochastic programming*, Annals of Operations Research **100** (2000), pp. 55–84. [6]

[103] R. SCHULTZ, *Stochastic programming with integer variables*, Mathematical Programming **97** (2003), pp. 285–309. [5]

[104] R. SCHULTZ AND S. TIEDEMANN, *Risk aversion via excess probabilities in stochastic programs with mixed-integer recourse*, SIAM Journal on Optimization **14** (2003), pp. 115–138.   [5, 7]

[105] R. SCHULTZ AND S. TIEDEMANN, *Conditional value-at-risk in stochastic programs with mixed-integer recourse*, Mathematical Programming **105** (2006), pp. 365–386.   [5, 7]

[106] M. SHAHIDEPOUR AND H. YAMIN, *Market Operations in Electric Power Systems: Forecasting, Scheduling, and Risk Management*, John Wiley and Sons, New York, 2002.   [33]

[107] G. B. SHEBLÉ, *Computational Auction Mechanisms for Restructured Power Industry Operation*, Kluwer Academic Publishers, Norwell, MA, USA, 1999.   [33]

[108] M. C. STEINBACH, *Tree-sparse convex programs*, Mathematical methods of operations research **56** (2002), pp. 347–376.   [1]

[109] R. TIEDEMANN AND C. FÜNFGELD, *Die Repräsentativen VDEW-Lastprofile – Der Fahrplan* (2003).   [61]

[110] S. TIEDEMANN, *Risk Measures with Preselected Tolerance Levels in Two-Stage Stochastic Mixed-Integer Programming*, Cuvillier Verlag, Göttingen, 2005.   [7]

[111] M. H. VAN DER VLERK, *Stochastic Programming Bibliography*, http://mally.eco.rug.nl/spbib.html.   [-]

[112] R. M. VAN SLYKE AND R. J.-B. WETS, *L-shaped linear programs with applications to optimal control and stochastic linear programming*, SIAM Journal on Applied Mathematics **17** (1969), pp. 638–663.   [26, 50]

[113] H. R. VARIAN, *Microeconomic Analysis*, Norton, New York, 1992.   [7]

[114] J. VON NEUMANN AND O. MORGENSTERN, *Theory of Games and Economic Behaviour*, Princeton University Press, Princeton, 1953.   [7]

[115] D. W. WALKUP AND R. J.-B. WETS, *Lifting projections of convex polyhedra*, Pacific Journal of Mathematics **28** (1969), pp. 465–475.   [6]

[116] G. A. WHITMORE AND E. FINDLAY, M. C., *Stochastic Dominance: An Approach to Decision Making under Risk*, D. C. Heath, Lexington, MA, 1978.   [9]

[117] W. WHITT, *Stochastic comparisons for non-markov processes*, Mathematics of Operations Research **11** (1986), pp. 609–618.   [8]

Symbol Index

$\subset$ $A \subset B :\Leftrightarrow x \in A \Rightarrow x \in B$, page 2

Ω Abstract set of elementary events, page 2

$\mathscr{F}$ σ-algebra/field on Ω, page 2

$\mathbb{P}$ Probability measure on $\mathscr{F}$, page 2

$\mathscr{B}^s$ The smallest σ-algebra in $\mathbb{R}^s$ containing all open subsets of $\mathbb{R}^s$, page 2

$\mathbb{E}$ Expected value operator on $\{X \mid X : \Omega \to \mathbb{R} \cup \{\pm\infty\}, \mathscr{F} - \overline{\mathscr{B}}\text{-measurable}\}$: $\mathbb{E}(X) := \int_\Omega X \, d\mathbb{P}$ (if the integral exists), page 2

$\oplus$ Minkowski addition of sets, $A \oplus B := \{a + b : a \in A \wedge b \in B\}$, page 6

$\mathbb{P}_X$ Also $X(\mathbb{P})$, image measure of $\mathbb{P}$ under X. $\mathbb{P}_X := \mathbb{P} \circ X^{-1}$, page 8

$\mathbb{1}_A$ Characteristic function defined on a set Ω that indicates membership of an element in a subset A of Ω, being 1 iff $x \in A$ and 0 otherwise, page 8

F_X Cumulative distribution function of the random variable X. $F_X(t) := \mathbb{P}\{X \leq t\} := \mathbb{P}[X^{-1}((-\infty, t])] = \mathbb{P}_X((-\infty, t])$, page 9

$(.)_+$ The positive part; $(.)_+ := \max\{., 0\}$, page 9

2^M The power set of M, page 14

$\overline{A}$ Let A be a subset of B. $\overline{A} := \{b \in B : b \notin A\}$, page 17

$\xrightarrow{\mathscr{D}}$ We say a sequence of random variables X_n converges in distribution to X, and write $X_n \xrightarrow{\mathscr{D}} X$, iff $\mathbb{P} \circ X_n^{-1} \xrightarrow{w} \mathbb{P} \circ X^{-1}$, page 20

conv   For $A \subset \mathbb{R}^n$, conv(A) denotes the convex hull of A. This is the smallest convex superset containing A, page 53

$\partial f(x_0)$ A vector $a \in \mathbb{R}^n$ is a subgradient of the convex function $f : \mathbb{R}^n \to \mathbb{R} \cup \{\pm\infty\}$ in x_0 iff $f(x) \geq f(x_0) + a^\top(x - x_0) \, \forall x \in \mathbb{R}^n$. $\partial f(x_0)$, called the subdifferential of f in x_0 is the set of all subgradients of f in x_0, page 53